Collins *Advanced Sciences*

Beha Populations

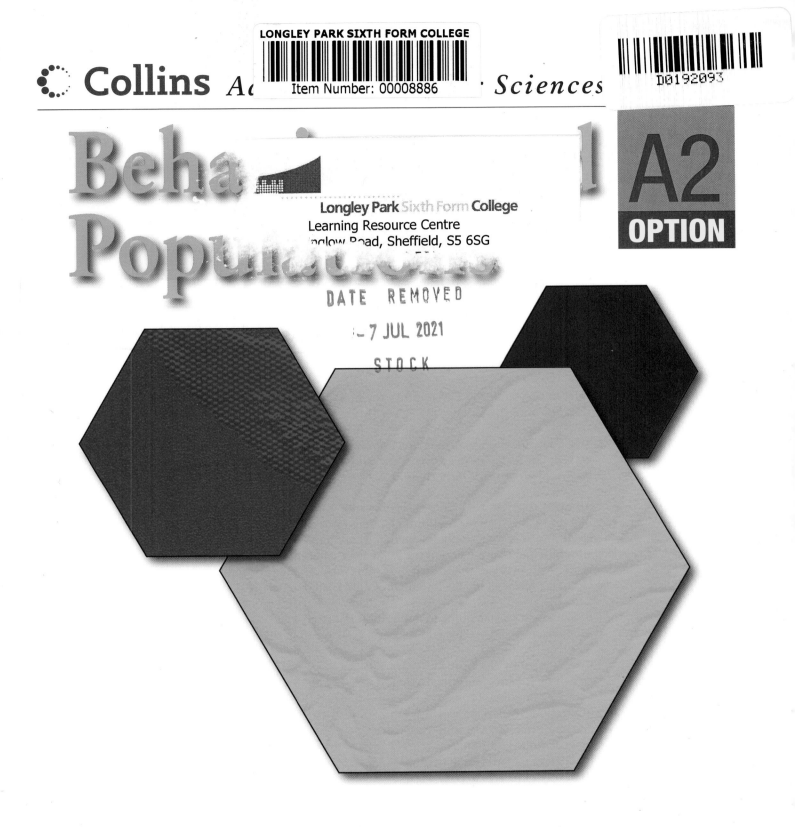

A2 OPTION

Series Editor: Mike Bailey

Pete Murray

Nick Owens

This book has been designed to support AQA Biology specification B.
It contains some material which has been added in order to clarify the
specification. The examination will be limited to material set out in the
specification document.

Published by HarperCollins*Publishers* Limited
77–85 Fulham Palace Road
Hammersmith
London
W6 8JB

Browse the complete Collins catalogue at
www.collinseducation.com

First published 2001
Reprinted 2001, 2003, 2004
ISBN 0 00 327743 7

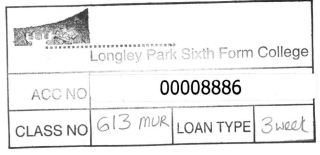

Peter Murray and Nick Owens assert the moral right to be identified as
the authors of this work

This book draws on some sections from *People, Population and Behaviour*
by Peter Murray and Nick Owens, and *Microbes, Medicine and Biotechnology*
by Ken Mannion and Terry Hudson.

British Library Cataloguing in Publication Data
A catalogue record for this publication is available from the British Library

Cover design by Chi Leung
Design by Jordan Publishing Design
Edited by Penelope Lyons
Illustrations by Barking Dog Art, Jerry Fowler, Peter Harper
Picture research by Caroline Thompson
Index by Julie Rimington
Production by Kathryn Botterill
Commissioned by Martin Davies
Project edited by Simon Gerratt

Printed and bound in Great Britain by Martins the Printers, Berwick upon Tweed

The publisher wishes to thank the Assessment and Qualifications Alliance
for permission to reproduce examination questions.

You might also like to visit:
www.harpercollins.co.uk
The book lover's website

CONTENTS

Acknowledgements

Text and diagrams reproduced by kind permission of:
Nature, Bird Study, Nelson & Sons Ltd., Cambridge University Press, McGraw-Hill Publishing Co., Hodder & Stoughton Ltd., Phaidon Press Ltd., Oxford University Press, *Behaviour*, Freeman & Co. Ltd., *Ecology*, Mosby-Wolfe, Pitman Publishing, *Clinical Review of Gynaecology*, Bourn Hall Clinic, Science Teachers Association of Western Australia, United Nations, Population Reference Bureau, Hutchinson, *Biological Sciences Review* Heinemann Educational Publishers.

Every effort has been made to contact the holders of copyright material, but if any have been inadvertently overlooked the publishers will be pleased to make the necessary arrangements at the first opportunity.

The publishers would like to thank the following for permission to reproduce photographs:
(T = Top, B = Bottom, C = Centre, L= Left, R = Right):

Ace Photoagency/Mugshots, 84C, Zephyr Pictures, 88CL;
BBC Natural History Unit/C O'Reilly, 19;
Biophoto Associates, 90;
Bruce Coleman Ltd/H Reinhard, 7, 11, G Cubitt, 12, A Purcell, 17R, Dr S Prato, 21;
Corbis/R Landau, 62, C & J Lenars, 16R;
Format Photographers/Judy Harrison, 53;
GettyOne Stone, 74;
Sally & Richard Greenhill, 72;

Robert Harding Picture Library/J Pottage, 16L, S Sassoon, 58;
Peter Murray, 50, 51;
NHPA/B Beehler, 17L;
Nick Owens, 22;
'PA' Photos, 6;
SPRI Picture Library, 86;
Science Photo Library, 28, 29, 39, 42, 43, 46, 56, 60, 68, 75, 77, 82, 84L&R, 88T, 92, 93, 94;
Southern Water plc, 69.

Front cover:
Images supplied by: GettyOne Stone (top left and centre), Science Photo Library (top right)

To the student

This book aims to make your study of advanced science successful and interesting. Science is constantly evolving and, wherever possible, modern issues and problems have been used to make your study stimulating and to encourage you to continue studying science after you complete your current course.

Using the book

Don't try to achieve too much in one reading session. Science is complex and some demanding ideas need to be supported with a lot of facts. Trying to take in too much at one time can make you lose sight of the most important ideas – all you see is a mass of information.

Each chapter starts by showing how the science you will learn is applied somewhere in the world. At other points in the chapter you may find more examples of the way the science you are covering is used. These detailed contexts are not needed for your examination but should help to strengthen your understanding of the subject.

The numbered questions in the main text allow you to check that you have understood what is being explained. These are all short and straightforward in style – there are no trick questions. Don't be tempted to pass over these questions, they will give you new insights into the work. Answers are given in the back of the book.

This book covers the content needed for the option module in AQA Specification B in Biology at A2-level: Module 8 – Behaviour and Populations. The Key Facts for each section summarise the information you will need in your examination. However, the examination will test your ability to apply these facts rather than simply to remember them. The main text in the book explains these facts. The case studies encourage you to apply them in new situations.

Words written in bold type appear in the glossary at the end of the book. If you don't know the meaning of one of these words check it out immediately – don't persevere, hoping all will become clear.

Past paper questions are included at the end of each chapter. These will help you to test yourself against the sorts of questions that will come up in your examination.

1 Patterns of behaviour

Is human aggression innate?

Every newspaper includes accounts of human aggression. Most people agree than human violence is undesirable and to be avoided if at all possible. Aggressive behaviour occurs in almost all animals, and one accepted definition of aggression is 'hostile behaviour directed towards causing physical injury to another individual'.

Konrad Lorenz referred to aggression as 'the fighting instinct in beast and man'. By using the word 'instinct', Lorenz implied that human aggression is innate and therefore inevitable. He saw competitive sport as an opportunity to release aggressive urges, which would otherwise be expressed in more damaging ways. However, there is no clear evidence that aggression in animals or humans *is* inevitable. In fact, football matches may induce aggression by offering the right stimuli in an emotionally charged situation.

1.1 Innate and learned behaviour

Innate behaviour (Table 1) depends on the genetic make-up of an animal and occurs without the need for any learning. For example, a newly emerged queen wasp accurately constructs a nest of hexagonal cells for her eggs, having never done so before, because she has inherited this ability.

Learned behaviour (Table 1) results from an animal's experience of the world. The animal's behaviour changes as a result of **stimuli** from its environment. For example, a young chimpanzee learns to use a stick tool to feed on termites by imitating the behaviour of other chimps. An animal's ability to learn depends on the structure of its nervous system, so learned behaviour, like innate behaviour, is influenced by genetic inheritance.

Birds

Migration

Birds that migrate at night, such as warblers and members of the thrush family, use the stars to find their way. A warbler called a blackcap (*Sylvia atricapilla*) has two populations in Germany. The population from central Germany migrates to winter in the UK; the other population from southern Germany migrates south-westward to winter in the Mediterranean. Blackcaps from these two populations were experimentally mated and the offspring tested for their migration direction (Fig. 1). The results showed that migration behaviour in blackcaps is inherited, or innate.

In geese, family groups migrate together to their wintering grounds. The brent goose (*Branta bernicla*) has three separate

Table 1 Differences between innate and learned behaviour	
Innate behaviour	**Learned behaviour**
inherited	not inherited
not changed by environment	changed by environment
inflexible	quickly adapts to new circumstances
similar in all members of a species	differs between members of a species

Fig. 1 Migration in blackcaps

**Blackcap females have a brown head.
The birds migrate at night.**

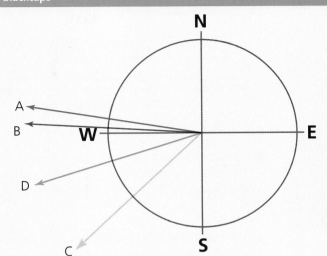

Blackcaps were tested in a glass-topped container during the migration season. They could see the night sky and fluttered towards the directions in which they would normally migrate. Each bird was tested 15–20 times. The offspring of UK-wintering birds migrate in the same direction as the parent birds. The offspring of crosses between the two populations migrate in a direction mid-way between those of the parent birds.

Blackcaps	Number of birds tested	Mean orientation in degrees
A adults wintering in the UK	18	279
B offspring of A	41	273
C adults wintering in the Mediterranean	49	227

Source: adapted from Berthold et al, *Nature*, **360**: 668–9; *Bird Study*, **42** (1995), 89–100

populations wintering in different parts of the British Isles and Europe. The young birds learn to find the traditional wintering areas by following their parents, and return to the same areas year after year (Table 2). Unlike blackcaps, geese live for a long time, and such learned behaviour allows them to adapt to changing conditions, but is not inherited by the next generation. In general, short-lived animals, such as insects, rely mostly on innate behaviour, while longer-lived animals, such as mammals and birds, depend more on learned behaviour.

Birdsong
Chaffinches have local dialects, showing that at least part of their song is learned by listening to other local birds. Chaffinch chicks reared in isolation in a sound-proof cage sing only a rough version of their song. They learn the full song by hearing other chaffinches, but they do not imitate the song of a blackbird. As with all animals, the genetic make-up of a chaffinch controls the development of its brain, and determines what it learns, how it learns, and when it learns most easily. This means that we cannot make a hard and fast distinction between innate and learned behaviour: all behaviour is a blend of both.

1 Wild canada geese migrate south every autumn from Canada to the USA. Canada geese bred in captivity and released into the wild in the UK have formed large populations that do not migrate.

a Explain why canada geese in Canada migrate whereas those in the UK do not.

b Explain why UK blackcaps bred in captivity migrate in the correct direction when released.

Table 2 Wintering areas of brent geese		
Population	**Breeding area**	**Wintering area**
Branta bernicla bernicla	N Siberia	SE England and NW France
Branta bernicla hrota	Greenland and Canada	Ireland
Branta bernicla hrota	Spitsbergen	Denmark and NE England

Mammals

Many young mammals show an even greater ability to learn than birds. A mammal's slow development and the high degree of parental care allow a long period of learning. By feeding the young on milk, a female mammal ensures a long period of close contact with her offspring, during which learning can take place. The capacity to learn means that a mammal's behaviour is highly flexible and adaptable to a wide range of environments.

Humans have a greater capacity to learn than any other species: only our reflex actions, such as blinking, or removing the hand from a hot object, are innate.

1.2 Taxes and kineses

Invertebrates, unlike mammals and birds, rely on innate behaviour for many aspects of their lives, such as escaping danger, finding a suitable habitat and locating food. They do this using three types of behaviour pattern:

● a type of orientation movement called a **taxis**;

● a type of orientation movement called a **kinesis**;

● a **reflex escape response** (see p. 9).

Taxis

Fly maggots use innate behaviour for survival. They avoid bright sunlight, which might harm them or make them visible to predators (Fig. 2). An animal performs a taxis (pl. taxes) when it moves towards or away from a stimulus, such as light, coming from a particular direction. In the case of the maggot it is **negative phototaxis** (taxis away from light). Adult flies have more protective pigments and usually move *towards* the light, which warms up their bodies (**positive phototaxis**).

Kinesis

Woodlice use innate behaviour to stay in a suitable environment. They live in damp places beneath logs and stones where they are not easily found by predators such as blackbirds and magpies, and are not likely to dry up. At high humidity levels, woodlice

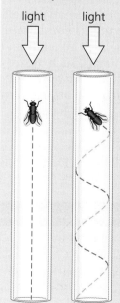

Fig. 2 Phototaxis

Negative phototaxis

7 6 5 4 ← light

3

2

1

Maggots have light receptors at each side of their heads. As they crawl, they turn their heads alternately to right and left, comparing the light intensity from each side. They always turn towards the darker side, and this takes them away from light.

↑ light

Positive phototaxis

light   light

Adult fruit flies move towards the light by keeping the light intensity the same in each eye. This can be demonstrated with a normally sighted fly and a fly that is blind in one eye. Each fly is placed in a glass tube with the light shining directly down from above. The normally sighted fly moves straight up the inside wall of the tube towards the light but the fly that is blind in one eye moves up the tube in a spiral. The attempt to keep the light intensity the same in each eye leads to the fly always having its blind eye turned towards the light.

remain inactive, but any slight drying of the environment is detected and they respond to the harmful stimulus by starting to move about.

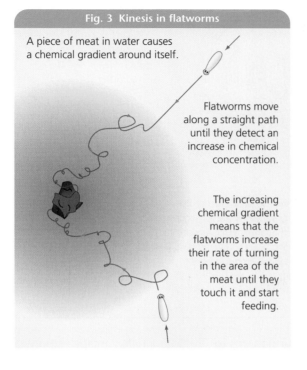

Fig. 3 Kinesis in flatworms

A piece of meat in water causes a chemical gradient around itself.

Flatworms move along a straight path until they detect an increase in chemical concentration.

The increasing chemical gradient means that the flatworms increase their rate of turning in the area of the meat until they touch it and start feeding.

Once it has started moving, a woodlouse keeps on moving until somewhere sufficiently moist is reached, then it slows down or stops completely. In a kinesis, unlike a taxis, the animal does not go in any particular direction with respect to the stimulus.

A slightly different kind of kinesis is seen in flatworms; they respond to chemicals given off by food in the water by increasing their rate of turning, though not in any set direction. If a piece of meat is placed in a pond, there can be several flatworms feeding on it within minutes (Fig. 3).

2 Twenty flatworms were placed in a dish of water in a well-lit room. Eventually, the flatworms became evenly spread around the dish. A light-proof cover was then placed over one half of the dish. One hour later, all the flatworms were in the dark half of the dish. Explain how the flatworms moved to and remained in the dark half of the dish if they were using:

a a taxis;

b a kinesis.

1.3 Reflex actions

A **reflex action** is a rapid, innate, automatic response to a stimulus.

Reflex escape response
Earthworms come to the surface of the ground on warm, damp nights to defecate or mate. They respond to the slightest vibration by retreating down their burrows. This is a reflex escape response, and it helps them to avoid being taken by a predator such as a shrew or hedgehog (Fig. 4, overleaf).

Complex innate behaviour in invertebrates
It would be a mistake to think that all innate behaviour is simple. For example, honey bees (*Apis mellifera*) use 'waggle dances' to communicate the direction and distance of a food source to other bees.

Complex innate behaviour in mammals
In mammals too, simple reflexes may be linked together to produce complex sequences of behaviour. For example, the innate stretch reflexes in the leg muscles are integrated into a functional pattern when an infant learns to walk. Similarly, several reflexes of mother and

baby operate together when a mother suckles her infant. First, the baby turns its head, with open mouth, towards the side where its cheek is touched by the breast, helping it to seek the nipple. This is known as the *rooting reflex*. Contact with the nipple initiates the baby's *sucking reflex*, causing the mother's nipple to evert into the baby's mouth, and this in turn stimulates nerve endings in the mother's skin. Nerve impulses from the **sensory nerve endings** travel from the skin to the mother's brain, causing the reflex release of the **hormone** oxytocin from the pituitary gland. Oxytocin then travels in the blood stream to the mammary glands causing the release of milk within a few seconds. In this *let-down reflex*, smooth muscle cells surrounding the alveoli in the mammary glands contract to squeeze milk into the sinuses surrounding the nipple. The baby is able to suck milk from these sinuses. The sucking by the baby therefore helps to regulate the output of milk from the mother. Breast-feeding also helps the mother and infant to form an affectionate bond.

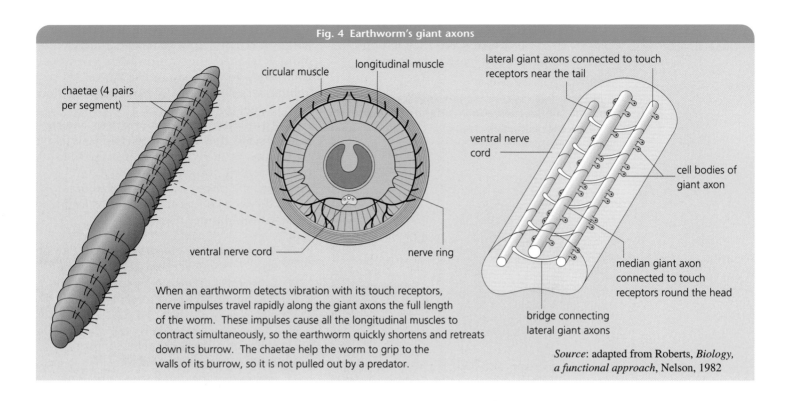

Fig. 4 Earthworm's giant axons

chaetae (4 pairs per segment)

circular muscle

longitudinal muscle

lateral giant axons connected to touch receptors near the tail

ventral nerve cord

cell bodies of giant axon

ventral nerve cord

nerve ring

median giant axon connected to touch receptors round the head

bridge connecting lateral giant axons

When an earthworm detects vibration with its touch receptors, nerve impulses travel rapidly along the giant axons the full length of the worm. These impulses cause all the longitudinal muscles to contract simultaneously, so the earthworm quickly shortens and retreats down its burrow. The chaetae help the worm to grip to the walls of its burrow, so it is not pulled out by a predator.

Source: adapted from Roberts, *Biology, a functional approach*, Nelson, 1982

KEY FACTS

- Invertebrates rely on innate behaviour patterns to find food and safety.

- In a taxis, the animal detects the direction of a stimulus, such as light, and moves towards or away from it.

- In a kinesis, the animal moves in a straight line until it meets conditions resembling those it needs. It then responds by either slowing down or increasing its rate of turning until the right conditions are met.

- In a reflex escape response, invertebrates move rapidly away from a stimulus that indicates immediate danger – for instance, an approaching predator.

- In humans and other mammals, a series of simple reflexes can link together to form a complex behavioural sequence.

1.4 Modified reflexes

Control of the bladder sphincter muscles requires us to modify the unconscious reflex of emptying the bladder. The stimulus of a full bladder causes the sphincter muscles around the base of the urethra to relax, allowing urine to be released. The conscious control of this reflex is an example of a modified reflex. It requires control of the sphincter muscles of the urethra, which are under the control of the autonomic nervous system (ANS). The ANS is the part of the nervous system that controls the actions of muscles not attached to the skeleton, such as those involved with heart-beat, blood pressure and gut movements. These muscles are not generally under voluntary control.

Most children learn to control this reflex even when they are asleep so that they do not wet their bed, but some children have difficulty in achieving this. These children can be helped using an alarm system which automatically wakes the child as soon as it starts to urinate. The child learns to associate the feeling of needing to urinate with the alarm bell and waking up. After a period of training the child is able to wake up before it starts to wet the bed. This is an example of a conditioned reflex (p. 13).

1.5 Learned behaviour

Learning can be defined as 'a change in behaviour caused by experience'. When an animal has learned something, its response to a particular stimulus changes. The change in behaviour is often permanent and not just a result of fatigue. Examples of learning which happen fairly rapidly include **habituation** and **imprinting**.

Habituation

Models of hawks or owls have frequently been used to deter small birds from flying into plate glass windows, eating crops or damaging trees and buildings. Playing the recorded distress calls of birds has also been tried. None of these methods works for very long, because the birds become habituated to them. Habituation is considered to be a type of learning because the response to a stimulus changes; the birds learn that the model or noise is not a sign of danger and cease to be deterred by it.

A scarecrow ceases to be effective when birds have become habituated to it.

Habituation in babies

Habituation occurs regularly in the lives of all animals, and prevents them from wasting time and energy responding to harmless stimuli. The figure below shows the startle reaction in babies.

Startle reaction in babies

when a repeating tone is sounded, the baby stops sucking and makes a 'startle reaction' to the noise

after 9 further bursts of repeating tone, the baby does not respond to the sound; it has habituated to the sound

startle reaction reappears when a different sound is made

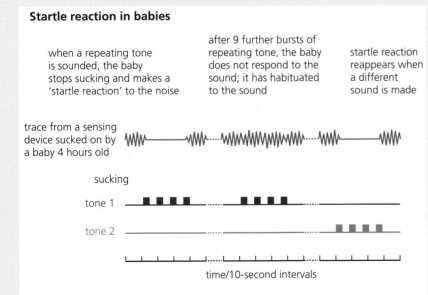

Source: adapted from Blakemore, *Mechanics of the Mind*, Cambridge University Press, 1977

1 Study the figure.

  a Explain why the startle reaction of a baby is a reflex action.

  b Explain why the baby ceasing to respond to the repeated sound is considered to be a type of learning (called habituation).

  c Explain why the startle reaction reappears when a different sound is made.

  d Explain how you could use this behaviour to investigate the ability of a baby to discriminate between different notes.

  e Suggest how habituation to sounds aids survival in human babies.

11

Imprinting

Chicks, goslings and ducklings flee the nest a few hours after hatching. They follow the first moving thing they see, which is usually a parent bird. After that, they only follow objects that look like the first object followed (Fig. 5). This learning is called imprinting. Imprinting ensures the young birds follow their parents, so the brood keeps together and avoids danger. There is a **sensitive period** when imprinting is extremely likely, but if it does not happen then, it may not happen at all. Imprinting in birds can occur to a wide range of objects that vary in size from a matchbox to a human. Round and conspicuous objects are followed more readily.

Adult birds often show courtship and reproductive behaviour towards the object of imprinting. This is known as **sexual imprinting**; it shows that they learn the physical characteristics of their own species during the imprinting process. Imprinting is also thought to be important in learning the characteristics of close kin, thereby helping to avoid inbreeding. In humans, incest is prohibited by an incest 'taboo'. This was once thought to be a learned cultural adaptation. However, evidence from unrelated children reared together (for example, in a kibbutz) shows that they rarely, if ever, choose each other as sexual partners. They treat each other as brothers and sisters. A process of negative imprinting apparently occurs up to the age of 6 or 7 years. During this period, a child learns which individuals to avoid as sexual partners later in life.

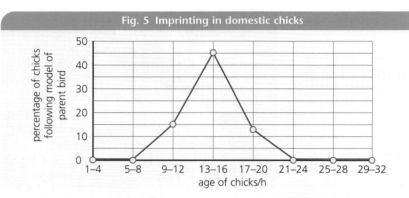

Fig. 5 Imprinting in domestic chicks

At first, the chicks' feathers are drying and they are not able to walk.
If the chicks are left for over 24 hours, they will not follow a moving object at all.

Source: data adapted from Hinde, *Animal Behaviour*, McGraw-Hill, 1966

APPLICATION

Saving the takahe

The New Zealand takahe is close to extinction. In an attempt to save the species, takahes are reared in captivity for later release into the wild. Dummy mother models are used to represent a parent bird. All feeding of the chicks is done using a glove puppet that looks like a takahe's head, and the young takahes are not allowed to see any view of the person feeding them.

These takahe chicks with dummy mother model are about to feed from a dummy parent beak.

1 Explain the consequences after release of the takahes of *not* using a glove puppet when feeding young takahes:

   a during the sensitive period;

   b after the sensitive period.

2 The models used for rearing takahes do not look precisely like real takahes, which vary in size and appearance.

   a Suggest why using an exact mimic of a takahe is not essential for this project.

   b Explain why inbreeding avoidance is important in re-establishing a viable takahe population.

KEY FACTS

■ Learning can be defined as a change in behaviour caused by the experience of particular stimuli. The change must not be caused by fatigue alone.

■ Learning can be as simple as ceasing to respond to a stimulus that caused a response before. This is called habituation.

■ Some birds follow the first moving object they see after hatching. Thereafter, they follow that object and no other. This is called imprinting. As adults, some birds show courtship and reproductive behaviour towards the imprinting object. This is called sexual imprinting.

Conditioning

The pets sharing our homes are quick to learn where food is to be found. The sound of a tin being opened brings a cat or dog into the kitchen very quickly, because the animal has learned to **associate** the sound of the tin opener with food. This type of learning is called **conditioning**. There are two types of conditioning: **classical conditioning** and **operant conditioning**.

Classical conditioning

Conditioning interested the physiologist Pavlov, who worked in Russia at the end of the 19th century. Pavlov wanted to test the effects of different types of food on the salivation reflex. He collected and measured the amount of saliva produced by the dogs when food was placed in their mouths. However, he noticed that his dogs started to salivate as soon as they heard his approaching footsteps, before he had given them any food. Pavlov then found that by giving the dogs a reward of food, he could train them to salivate to a stimulus such as a flashing light or a ringing bell. Pavlov called this type of learning, in which the usual stimulus (food) is replace by a new one (light or bell), a conditioned reflex (Fig. 6). Because this was the first type of conditioning to be described, it is now known as classical conditioning.

Operant conditioning

Pavlov's work encouraged others to study the way animals learn. Skinner developed a special apparatus (the Skinner box) for training an animal, usually a rat. By pressing a lever in the box, the rat gains a reward of a food pellet. At first, the rat presses the lever just by accident, but soon learns to associate lever pressing with a reward. This is called operant conditioning, because the animal is rewarded for an operation (movement) that it does naturally from time to time. The food reward makes it more likely that the rat will press the lever again; in other words the food is a positive **reinforcement** of the behaviour.

All kinds of movement can be reinforced in operant conditioning. For example, if a pigeon is rewarded with food every time it preens its feathers, the rate of preening quickly increases. **Reinforcers** need not be food: monkeys learn for the reward of seeing another monkey, chickens for the reward of straw to nest in, and dogs for the reward of attention from their owner.

Animals can also be conditioned using punishment. A stern telling-off works as a punishment and prevents a dog from jumping up to greet people, so long as the training is done consistently for a few months.

Conditioning is part of the natural lives of animals. For example, caterpillars of the mullein moth have a nasty taste and bright orange spots. The bad taste works as a punishment for any bird that attempts to eat the caterpillar, and the bird learns to associate the orange spots with a bad taste. This is an example of **warning coloration**.

Learned human behaviour

Learning plays a major part in our everyday lives. For instance, conditioning is used by advertisers. Some advertisements try to familiarise us with a particular sign or logo, so that we are more likely to select it from a range of similar logos. More commonly, advertisements try to make us associate their product with being attractive, successful or intelligent. We are encouraged to think that buying the product will give us these qualities.

Parents often use money to reward children for good behaviour. They also use various forms of punishment. This is rather like operant conditioning in rats, but humans can often see the motives behind other peoples' behaviour and resist being 'conditioned'. Sociologists and psychologists study the effects of reward and punishment on human behaviour, for example the effects of prison sentences as a deterrent against crime. While most people agree that too much punishment can make a child withdrawn and unresponsive, some think that too many rewards can encourage selfishness. Skilled parents and teachers are good at giving the right amount of control and encouragement to each child.

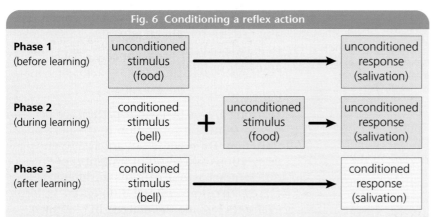

Fig. 6 Conditioning a reflex action				
Phase 1 (before learning)	unconditioned stimulus (food)	→		unconditioned response (salivation)
Phase 2 (during learning)	conditioned stimulus (bell) **+**	unconditioned stimulus (food)	→	unconditioned response (salivation)
Phase 3 (after learning)	conditioned stimulus (bell)	→		conditioned response (salivation)

Source: adapted from Gross, *The Science of Mind and Behaviour*, Hodder and Stoughton, 1980

- In conditioning, animals learn to associate one thing with another. Conditioning usually needs a reward to reinforce the desired behaviour.

- In classical conditioning, the usual stimulus causing a reflex action is replaced by a new stimulus.

- In operant conditioning, a natural behaviour of the animal is reinforced so that it happens more often.

- Human behaviour is influenced partly by the rewards and punishments given out by society.

Social and cultural aspects of learned behaviour

The most extensive learning occurs in social animals, which can often recognise each other as individuals.

Social learning

Geese and swans recognise members of their own families among the flock. Bewick's swans each have a distinctive pattern of yellow and black on the beak (Fig. 7). Young Bewick's from previous years sometimes join their

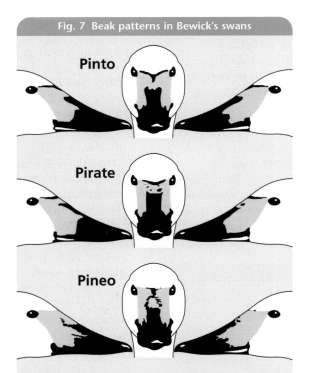

Fig. 7 Beak patterns in Bewick's swans

Pinto

Pirate

Pineo

'We quickly realised that the patterns of black and yellow on Bewick's swans' bills were infinitely variable. By drawing them in front and side view we could record the different patterns and give each swan a number and a name. We have now recorded the face patterns of 3,700 Bewick's swans'. PETER SCOTT

Source: adapted from Scott, *Observations of Wildlife*, Phaidon, 1980

parents and the current year's cygnets, making winter family groups of up to 17. This may be because traditional knowledge is vital in times of cold weather, when memory of alternative feeding grounds is needed.

The most lengthy period of immaturity occurs in **primates** (monkeys and apes). Play behaviour helps young primates to develop social responses, so that they are neither too aggressive nor too submissive towards others. In many primates there is a **dominance hierarchy**, often with one overall dominant animal, sometimes called the alpha male. The second-ranking animal is dominant to all except the alpha male and so on down the hierarchy. In some primates, such as baboons and rhesus monkeys, there are separate dominance hierarchies among males and females. A hierarchy helps to avoid excessive conflict over resources as the more dominant animal always takes priority. The skills needed to establish a high rank may be developed during play. Human play, including formal games, helps us to gain confidence and learn to communicate and co-operate. It also helps us to learn the skills, strengths and weaknesses of ourselves and others.

Competing successfully for food or a mate in a social group often means outwitting other group members. Primates may predict what others will do under certain conditions and use this to gain an advantage. Learning to make predictions successfully takes time and a lot of intelligence.

Culture

Learning allows behaviour patterns to be passed from one generation to the next by imitation. Differences can arise between populations as a result of imitation of different behaviour patterns in different places. These are cultural differences. In birds, examples include the dialects in chaffinch songs and the migration patterns of geese and swans. In chimpanzees, different populations have developed particular tool-using skills, such as fishing for termites with a stick.

Culture is a defining characteristic of human behaviour. We accumulate modifications and improvements to our culture by making better or different tools, machines, clothes, music and art each generation. This cultural evolution is greatly facilitated by the use of written and spoken language, and more recently by computers.

KEY FACTS

- Birds and mammals usually have slow development and a long period of parental care.

- Extended parental care provides time for exploration and play, allowing a long time for learning about the environment and about other members of the species. This gives the young a greater chance of survival and breeding success.

- Learning is especially important for primates. Primates living in groups must learn the identity and behaviour of other group members.

- Learning by imitation allows cultural evolution to occur.

EXAMINATION QUESTIONS

1 Frogs jump out of the way of a large object approaching them. The drawings show the results obtained when a large object was moved towards a frog from different directions. The arrows indicate the number of jumps in a particular direction; the longer the arrow, the greater the number of jumps.

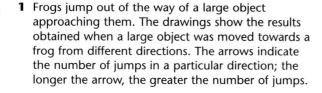

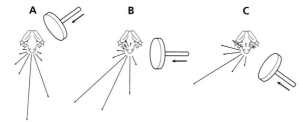

a Describe the relationship between the direction of movement of the object and the reaction of the frog. (2)

b Explain how the data support the idea that this reaction could be considered to be a taxis. (2)

AQA/NEAB BY04 March 1998 Q5

2

a Chimpanzees learned to put tokens into a machine that gave them a grape in return for each token. Explain why this behaviour might be described as an example of operant conditioning. (3)

b Explain what is meant by habituation. (2)

AQA/NEAB BY04 March 1998 Q6

3 Young chaffinches that have been hand-reared from eggs and isolated from other chaffinches produce songs of the same length and containing the same notes as wild chaffinches, but the phrasing is abnormal. These birds never develop a normal song pattern, even if they are later exposed to songs of wild chaffinches.

Young chaffinches caught just after leaving the nest and isolated from other chaffinches develop an almost normal song pattern. If these chaffinches are exposed to the songs of adult chaffinches when they are eleven months old they develop a completely normal song pattern, similar to the adult birds around them.

a Use the information in the passage to describe and illustrate the main features of:
   i)   innate behaviour;
   ii) imprinting. (8)

b Suggest why it is important for chaffinches to develop a normal song pattern. (4)

AQA/NEAB BY04 February 1997 Q7

2 Reproductive behaviour

What qualities do we look for in a partner?

Studies by Professor Dunbar of Liverpool University of lonely hearts advertisements in newspapers show that women tend to seek wealthy men a little older than themselves, while men prefer younger, physically attractive women. Men mentioned points about their attractiveness less often than women, but boasted material wealth more often. Men only stated their height if they were quite tall. This analysis suggests that we know what we want in a partner and also have some idea about what the opposite sex is seeking in us.

Who makes the ideal mate?

2.1 Courtship

Courtship is behaviour used to attract a mate. Courtship enables animals to:

- recognise their own species;
- approach each other closely without triggering aggression;
- choose a strong and healthy mate;
- form a pair bond and synchronise breeding behaviour.

Species recognition

Every animal species has a different courtship display that helps it to recognise and attract a member of its own species of the opposite sex. Species recognition is especially important in closely-related animals where a mistake could easily be made. Many kinds of ducks pair up on the wintering grounds where several similar species flock together. Each duck species has a different iridescent flash of colour in a bar on the wing, by which it can be recognised. Female ducks need to be camouflaged when sitting on their eggs, so

their colour is dull and the wing bar is only visible in flight. Male ducks do not incubate the eggs and many have evolved spectacular plumage and courtship displays. However, male ducks on isolated islands tend to be dull in colour like the female. This is probably because only one species usually colonises each island, so species recognition is not a problem. Camouflage is then more important than having bright colours.

Insects such as butterflies also use brightly coloured patterns as recognition signals, while fireflies use characteristically timed sequences of flashes at night. Species recognition can also be by sound or smell: the calls of frogs and crickets differ between each species, and many insects use chemical signals, called pheromones, for recognition (p. 21).

Avoiding aggression

Most animals have some fear of each other and maintain a small space between themselves and their neighbours. This is

The bird of paradise shows off its tail in courtship display.

Cuttlefish use rippling colour changes to attract a mate.

Female animals often select males with the most noticeable displays. Male swallows (*Hirundo rustica*) with the longest tail streamers are the most successful at attracting mates. This is **inter-sexual selection**, and it leads to the evolution of bigger and better courtship displays and physical adornments, such as the peacock's tail and the elk's antlers. There are two interpretations of inter-sexual selection. One is that animals select mates purely for their physical attractiveness. Almost any physical feature may be selected for, but once the selection begins, adornments tend to become more and more extreme. Animals acquiring the most attractive mates will in turn have offspring which are good at attracting the opposite sex, and their genes will be passed on.

An alternative interpretation is that long tails and big antlers are a signal that the male possesses good genes. These are, in a sense, handicaps to the male, as they use up energy and make him more susceptible to predators. Genetically unfit males will not be able to produce bright colours or large adornments. A male which can survive and flourish despite such handicaps must be of good genetic quality. It has been suggested that human activities such as body-piercing and the use of drugs are forms of self-handicapping; they advertise to the opposite sex that a person is fit enough to overcome the handicap, and is therefore a good potential mate.

Polygamous species, in which one partner has several mates, show the greatest degree of physical differences between the sexes (**sexual dimorphism**). Examples include southern elephant seals (*Mirounga leonina*) and African lions (*Felis leo*), in which one male monopolises several females. Large male body size arises through selection for the males' ability to fight with other males for access to females. This is **intra-sexual selection**.

There is considerable sexual dimorphism between men and women in size and body shape. Some of these differences may have arisen through natural selection rather than sexual selection. For example, among our hunter–gatherer ancestors, men needed strength and stamina for hunting with spears. However, sexual selection may be responsible for our loss of hair, our facial features and even our large brains. Larger-brained men or women are likely to be better at outwitting their rivals and showing more subtle courtship behaviour.

known as **individual space**; it provides some safety from aggression, and reduces the risk of infection (Chap. 7). Courtship enables males and females to enter each other's individual space without triggering aggression. This is usually a gradual process in which the members of the pair get to know each other.

In black-headed gulls, the pair first stand side by side, head to tail, with their dark faces – a signal of aggression – turned away from each other. They then make 'choking' movements, imitating the action of a parent gull regurgitating food for its chick. This seems to reassure the partner, and leads eventually to pair formation and mating.

Choosing a strong and healthy mate

When animals select a mate and start to breed, they are committing themselves to a major investment of time and energy. The female invests energy in her eggs, and the male invests energy in courtship and defending the territory. One or both partners may help to feed and protect the young. This **parental investment** often differs between the male and the female of a species.

Animals can improve their reproductive success by selecting a mate which will make more investment in the offspring. Some female birds, such as terns, test out a potential mate by inviting him to feed her. If a male tern can deliver plenty of fish, the female will stay with him. This may help to explain why human females seek mates with good financial resources.

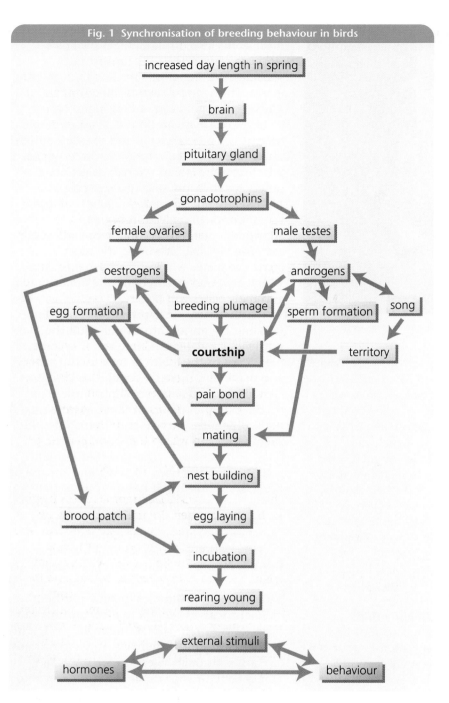

Fig. 1 Synchronisation of breeding behaviour in birds

Pair bond formation and synchronisation of breeding behaviour

In birds, such as starlings, increasing day-length in spring stimulates the brain. The pituitary gland responds by secreting **gonadotrophic hormones**, which cause the ovaries or testes to grow. Sex hormones from the ovaries and testes make the plumage more colourful, and cause the male to start singing and begin courtship behaviour. The male's courtship further stimulates the female's gonadotrophins and her ovaries respond by starting to produce eggs. Egg development and further courtship then encourage the female to mate. The whole sequence of laying, incubation and rearing young then follows, with each component triggering the next and having positive feedback effects on hormone levels.

During courtship, the male and female bird gradually develop a **pair bond** as their hormone concentrations increase and they begin breeding activities. The pair bond means that the members of the pair recognise each other as individuals and that they behave together in a co-ordinated way. There is a continuous interplay between environmental stimuli, hormones and behaviour which allows the pair to synchronise their breeding activities and dovetail their behaviour into a co-ordinated plan (Fig. 1). Pair bonds can last a lifetime in some bird species, including swans and geese, and in some mammals, such as gibbons. In some colonially breeding birds, such as black-headed gulls and sandwich terns, the nesting behaviour of the whole colony is synchronised. Because the eggs are all laid at about the same time, there is less opportunity for predators to take them.

KEY FACTS

■ Courtship enables potential mates to recognise their own species, avoid aggression, form a pair bond and synchronise their breeding behaviour.

■ Sexual selection can lead to the development of elaborate courtship displays and differences in appearance, size and behaviour between the sexes.

■ Animals often come into breeding condition in the spring, as a result of the stimulus of increased day-length.

■ Courtship behaviour can be analysed by looking at the sequence of its individual components. Each component is a response to a stimulus from the environment, or from a hormone, or from the partner's behaviour.

■ During courtship, a pair bond is formed allowing the male and female to coordinate breeding behaviour.

2.2 Territorial behaviour

A **territory** is defined as any defended area. Defending a territory requires a lot of energy. The benefits of territorial behaviour differ from species to species, but all can help to increase breeding success. Benefits include:

- maintenance of a food source;
- defence of a nest site;
- attracting a mate.

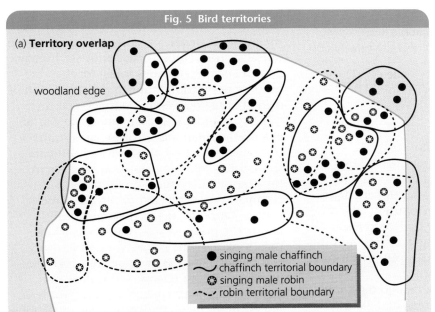

Fig. 5 Bird territories

(a) Territory overlap

woodland edge

- ● singing male chaffinch
- ～ chaffinch territorial boundary
- ✹ singing male robin
- ⌁ robin territorial boundary

Symbols within a territory are the same male bird heard singing on different days. Area of map = 60 000 m².

The overlap of chaffinch and robin territories shows that chaffinches and robins do not compete with each other for territory.

Source: adapted from Common Birds Census for Short Wood, Northamptonshire, 1995

(b) Territory size

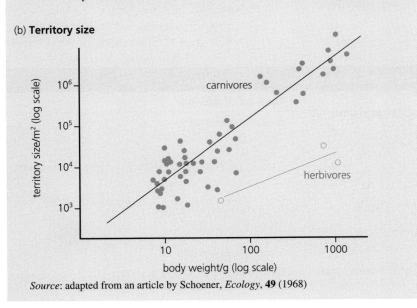

Source: adapted from an article by Schoener, *Ecology*, **49** (1968)

Defending a food source

Bird song can be a source of delight to humans, but to the birds concerned it is a means of claiming a territory, which preserves a food source for the adults and their young. Many birds, including wrens and chaffinches, defend a territory only during the breeding season. Others, such as robins, defend a territory during the winter as well. Usually, territories are only defended against members of the same species. Each member of a community sharing a habitat, such as a wood, occupies a different **ecological niche**, and competition between species is largely avoided (Fig. 5a). For example, robins usually feed by watching for moving prey on the ground, using a look-out perch, then hopping down to take the food. Chaffinches feed more in the tree canopy and take seeds as well as insects. As expected, larger species of birds usually have larger territories; a wren defends less than a hectare of woodland (a hectare is a square with sides 100 m), whereas a golden eagle defends an area of up to 90 km² of moorland and mountain. The bigger the bird, the more food it needs to rear its young. Carnivorous birds tend to have bigger territories than herbivorous birds (Fig. 5b). This is because less food energy is available at higher trophic levels.

Defending a nest site and attracting a mate

Skylarks often choose to nest in grassland or cornfields. As there are no trees in such habitats, skylarks sing in flight, high above the nest site. Female skylarks are attracted to males with good nesting sites in their territories. Once the best territories are occupied, other skylarks are forced to breed in poorer habitats where they are less successful.

Defending a nest site is even more important in cliff-nesting birds. In the Farne Islands, researchers have found that shags prefer to select the broader cliff ledges that are protected from the waves but are still close to the sea. The birds defend a small territory just around the nest site, but they feed on fish out to sea. In shags, the quality of the nest site is closely related to breeding success (Table 1, overleaf). The quality of the nest site was judged on a scale of 0–4 before nesting began, and breeding success was measured by the number of chicks fledged per pair.

23

Table 1 Nest site and breeding success		
Nest site quality	Mean number of chicks fledged per pair	*n* (no. of pairs)
0 (poor)	0.61	191
1	0.69	402
2	1.03	704
3	1.30	566
4 (good)	1.55	116

Source: adapted from Potts *et al.*, *Journal of Animal Ecology*, **49**, 1980

5 Study Table 1.

a What does it tell you about the relationship between nest site quality and breeding success?

b Suggest why shags only defend a small area around their nest site.

Territory in other animals

Territorial behaviour during the breeding season is common among mammals and birds, and is also seen in a few invertebrates such as dragonflies and octopuses (Table 2). All these animals are able to defend resources from other members of their species, and this helps them to attract a mate and/or rear more offspring.

In animals that do not defend territories, it may be more energy efficient to defend individual food patches instead. This happens if the location of a food supply is unpredictable, such as a moving shoal of fish or flower patches that bloom periodically, or if the food supply is simply too spread out to defend economically.

Human territories

Like many animals, we maintain a moving territory around us and feel uncomfortable if someone invades our individual space. When we are forced to be in close proximity, as in a crowded lift, we keep quiet, adopt a neutral posture and avoid eye contact. Courtship allows potential partners to break down this barrier and approach closely.

Humans often defend their living area. If we are discovered in someone else's house or garden, we are likely to be rebuked or even attacked. However, we may be invited in by the owner if we are a friend or if we introduce ourselves first. Unlike many animals, humans do not defend territories solely for the purpose of breeding, though the possession of a good home may help to attract a partner.

On a grander scale, humans go to war to defend their national territory against attack from another national power. Few animals show behaviour comparable to humans in this respect, although a group of male chimpanzees at the Gombe Stream National Park in Tanzania was seen systematically killing the males in a second group. They then took over the second group's territory and females. Such behaviour must increase breeding success. However, this does not mean that human territorial aggression is 'in the genes' or inevitable. As we saw in Chapter 1, the environment and learning are of great significance in controlling behaviour: we can modify our environment and our behaviour if we choose.

Table 2 Territorial defence		
Species	Territory	Method of defence
dragonfly (*Libellula depressa*)	pond	chasing other males
stickleback fish (*Gasterosteus aculeatus*)	area of riverbed around nest	chasing other males displaying red throat and spines
Canadian timber wolf (*Canis lupus*)	hunting area of the pack (10 000–12 000 km²)	scent marking using urine howling
domestic cats (*Felis felis*)	hunting and breeding area	mewing and fighting
	male's territory includes territory of several females and is up to 10 times bigger	scent marking by spraying urine rubbing scent glands on twigs and stems

APPLICATION **Breeding birds in woodland**

Year	No. of territories	
	Blue tit	Great tit
1	10	3
2	9	4
3	9	6
4	13	6
5	14	6
6	10	4
7	17	5
8	22	7
9	12	4
10	14	4
11	19	8
12	15	7
13	20	8
14	22	10
15	21	7
16	18	6
17	17	7

A study was made of great tits and blue tits breeding in a broad-leaved woodland in Northamptonshire over a period of 17 years. Blue tits feed largely in the tree canopy, while great tits feed more in the shrub layer and on the ground. A standard census method was used to estimate the number of territories of each species nesting in 10 hectares of woodland. The results are shown in the table on the left.

1 a Work out the mean and standard deviation for the number of territories in the study area for each species over the 17 years of the study.

b Use your figures from (a) to show if the number of blue tit territories was statistically significantly greater than the number of great tit territories, using a t-test by substituting in the formula:

$$t = \frac{\bar{x}_a - \bar{x}_b}{\sqrt{\dfrac{s_a^2}{n_a} + \dfrac{s_b^2}{n_b}}}$$

where

$\bar{x}_a$ = mean of first sample

$\bar{x}_b$ = mean of second sample

s_a = standard deviation of first sample

s_b = standard deviation of second sample

n_a = sample size of first sample

n_b = sample size of second sample.

Use the table for values of t on p. 27.

2 a Use your figures for the mean number of territories in 1(a) to work out the average territory size in hectares for the two species.

b Assuming the average mass of a blue tit is 10 g and a great tit is 20 g, explain why a great tit needs a larger breeding territory than a blue tit. State any further assumptions that you make.

c Suggest what would happen to great tit breeding success if:

(i) the number of great tits breeding in the wood increased;

(ii) the number of blue tits breeding in the wood increased.

d Comment on the relative sizes of great tit and blue tit territories compared with their relative masses.

KEY FACTS

- Some animals improve their breeding success by defending a territory. This helps them to maintain a reliable food supply, attract a mate or protect a good nest site.

- Territories are usually defended from members of the same species, since each species occupies a different ecological niche.

- Territories may be defended by visual displays, fighting, songs, sounds and smells.

- Breeding territories occur only in species where resources can be economically defended.

- Humans have territories, but these are often poorly defined and not closely tied to breeding success. Human territorial aggression is not an inevitable part of our behaviour.

1 Cockroaches are insects that live in cracks in walls, emerging to feed at night. In its natural environment the behaviour of a cockroach is influenced by contact between its body and objects under which it moves, and by the presence of a pheromone.

In order to investigate the relative importance of these two stimuli, a specially designed choice chamber was used. This was made of glass plates separated by spaces. The distance between the plates on side A was just sufficient to allow cockroaches to move between them. The diagram shows the apparatus with the sides removed to show the arrangement of the glass plates. The space between the top plates was sealed after the cockroaches had been added.

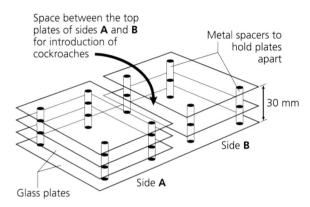

Space between the top plates of sides **A** and **B** for introduction of cockroaches

Metal spacers to hold plates apart

30 mm

Side **B**

Side **A**

Glass plates

A series of pheromone dilutions was made to treat the surfaces of side B of the chamber. For each dilution, 60 cockroaches were put into the chamber. The cockroaches were then left for 24 hours in total darkness before their distribution was recorded.

Concentration of pheromone/ arbitrary units	Number of cockroaches in side B after 24 hours
0	1
1	4
2	5
3	6
4	7
5	10
6	15
7	30
8	50

a i) What is a pheromone? (1)

ii) Describe the role of pheromones in influencing behaviour in nature. (3)

b i) What is the relationship between the concentration of pheromone and the number of cockroaches in side **B** after 24 hours? (1)

ii) Give one conclusion that can be made about the relative importance of body contact and pheromone concentration in determining the response of the cockroaches. (1)

iii) Suggest one way in which the behaviour shown by cockroaches in this experiment may help cockroaches to survive in their natural environment. (1)

AQA/NEAB BY04 February 1997 Q3

2 Robins have individual territories during the autumn and early winter, but from early in the new year pairs of birds begin to share the same territory, which they maintain throughout the spring and summer.

a Describe and explain the advantages of territorial behaviour, with reference to the behaviour of the robin throughout the year. (7)

b The territories are usually defended by song and displays, which often involve exhibiting the red breast as much as possible to any intruding robin.

Fighting is sometimes involved, especially when the territory is being established.

Suggest why robins usually defend their territories by song and display rather than fighting. (3)

c A robin shows aggressive behaviour towards models of adult robins and bunches of red feathers placed within its territory. However, models of young robins, which do not have a red breast, are not threatened.

What can be deduced from these observations about the factors controlling the aggressive behaviour in robins? (2)

AQA/NEAB BY04 June 1997 Q8

3 The flow chart shows how light affects the reproductive activity of some birds.

Increased day-length

↓

Receptors stimulated

↓

Hormones released from anterior lobe of pituitary gland

↓

Sex organs increase in size and activity. Courtship behaviour begins.

EXAMINATION QUESTIONS

a Suggest why day-length is a more appropriate stimulus to start courtship behaviour than temperature. (1)

b Give two reasons why courtship behaviour is important in birds. (2)

c Hens reared intensively for egg production are housed in large buildings without windows. Use the information in the flow chart to explain why, despite the expense, the lights are left on all the time. (2)

AQA/NEAB BY04 June 1996 Q7

REFERENCE MATERIAL

Chi-squared test

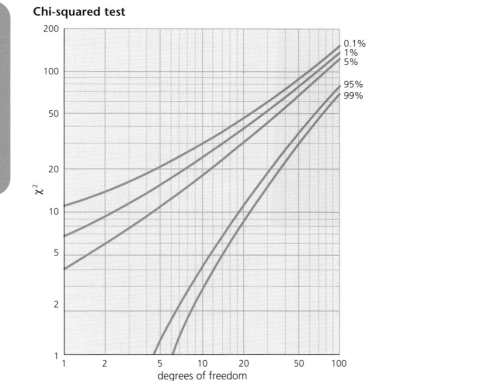

	Values of t				
	p%				
Degrees of freedom	**10**	**5**	**2**	**1**	**0.1**
10	1.81	2.23	2.76	3.17	4.58
11	1.80	2.20	2.72	3.11	4.44
12	1.78	2.18	2.68	3.05	4.32
13	1.77	2.16	2.65	3.01	4.22
14	1.76	2.14	2.62	2.98	4.14
15	1.75	2.13	2.60	2.95	4.07
20	1.72	2.09	2.53	2.85	3.85
30	1.70	2.04	2.46	2.75	3.64
40	1.68	2.02	2.42	2.70	3.55
50	1.68	2.01	2.40	2.68	3.50
60	1.68	2.00	2.39	2.66	3.46

3 Human reproduction

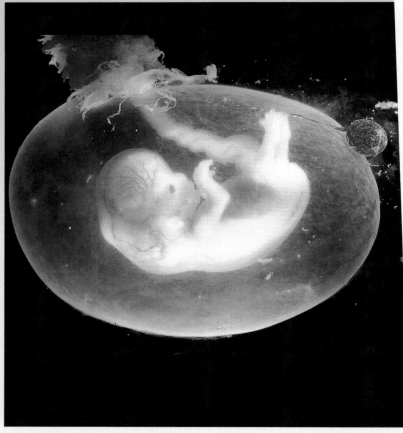

In ancient Egypt, diagnosing pregnancy involved the woman drinking a mixture of pulped watermelon and breast milk from a mother who had borne a son. Unfortunately, a positive result was indicated by a period of violent sickness. Another, less harrowing, idea of the ancient Egyptians was to pour women's urine over corn seeds to see if they would germinate. The first reliable biological tests were introduced just 60 years ago. It was discovered that when urine from pregnant women was injected into certain animals, it stimulated them to produce eggs. At first mice were used, but it took five days to produce a result. Rabbits were quicker and gave the result in 24–48 hours, but the Hogben Test, which used a female South African clawed toad, worked within just a few hours. The Hogben Test was carried out in some specialist laboratories until the 1960s. During the 1980s, reliable home pregnancy tests were developed, in which an immunological reaction was detected by a colour change. These tests enabled the detection of pregnancy from the first day of the missed period.

Source: adapted from Lesley Foster, Unipath Ltd. Pregnancy Testing. NCBE Newsletter Winter 1990

This human embryo is 7–8 weeks old. It is floating in an amniotic sac filled with amniotic fluid and is attached to the placenta by the umbilical cord.

Mammals have a reproductive cycle in which the female becomes receptive to mating every few weeks or months, sometimes at a particular time of year. A female mammal in mating condition is said to be in oestrus. For example, in red deer (*Cervus elephas*) there is a rutting season in the autumn when the stags compete for and mate with the hinds. The hinds then give birth the following spring when food is abundant. Humans are primates, the order of mammals that comprises monkeys and apes. Female primates have a unique reproductive cycle, known as the **menstrual cycle**, in which the lining of the uterus repeatedly thickens and breaks down.

3.1 The menstrual cycle

A woman's menstrual cycles begin in her early teens and finish with the **menopause**, at about the age of 50 years. The length of each cycle varies from woman to woman, and may be between 25 and 35 days, averaging 28 days. During each cycle:

- an egg is released from one of the woman's ovaries;
- the lining of her uterus (womb) is made ready to accept and support an embryo;

- if the egg is not fertilised, **menstruation** occurs, and the cycle begins again.

The uterus
The uterus consists of two layers of tissue:

- An outer layer, the **myometrium**, with smooth muscle used to give birth to the baby;

- An inner lining, the **endometrium**, which is shed and regenerated each cycle (Fig. 1).

Fig. 1 Human female reproductive system

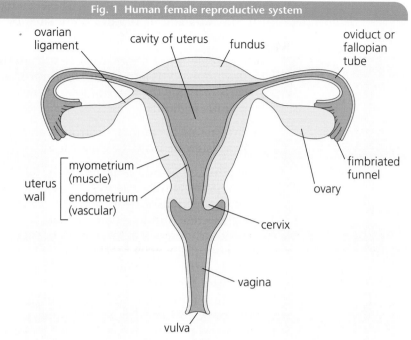

ovarian ligament • cavity of uterus • fundus • oviduct or fallopian tube • myometrium (muscle) • endometrium (vascular) • uterus wall • fimbriated funnel • ovary • cervix • vagina • vulva

Source: Adapted from Gadd, *Individuals and Populations*, Cambridge University Press, 1983

The ovaries

The ovaries contain **follicles** in which **ova** develop (Fig. 2). At birth there are thousands of **primordial follicles** in each of a female baby's ovaries. The ovum in these primordial follicles is not fully formed: it is halted at prophase I of meiosis, and is called a **primary oocyte**.

After puberty, a few of the primordial follicles develop and mature during each cycle. The cells surrounding the oocyte divide to form **granulosa cells**, which then secrete fluid into a cavity within the follicle. As the follicle grows and expands, the primary oocyte completes meiosis I, producing a **secondary oocyte** and a **polar body**. The secondary oocyte stops development at metaphase of meiosis II and will not become a haploid ovum until after penetration by a sperm. The polar body is the other nucleus produced by meiosis I and it develops no further. It remains in the small space around the secondary oocyte, leaving almost all the nutrients and cytoplasm in the oocyte (Fig. 3, overleaf).

1a Explain why almost all the nutrients are placed into one oocyte (leaving three polar bodies) rather than producing four equal-sized oocytes.

b Suggest why, in contrast, four sperms are made from each meiotic division in the testes, sharing the nutrients equally.

Fig. 2 Ovarian cycle

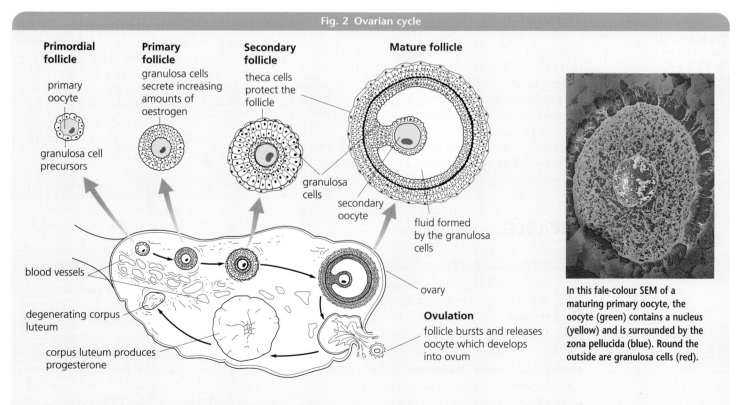

Primordial follicle — primary oocyte, granulosa cell precursors
Primary follicle — granulosa cells secrete increasing amounts of oestrogen
Secondary follicle — theca cells protect the follicle, granulosa cells
Mature follicle — secondary oocyte, fluid formed by the granulosa cells

blood vessels • degenerating corpus luteum • corpus luteum produces progesterone • ovary

Ovulation — follicle bursts and releases oocyte which develops into ovum

In this fale-colour SEM of a maturing primary oocyte, the oocyte (green) contains a nucleus (yellow) and is surrounded by the zona pellucida (blue). Round the outside are granulosa cells (red).

Source: adapted from Berne and Levy, *Principles of Physiology*, Wolfe, 1990

29

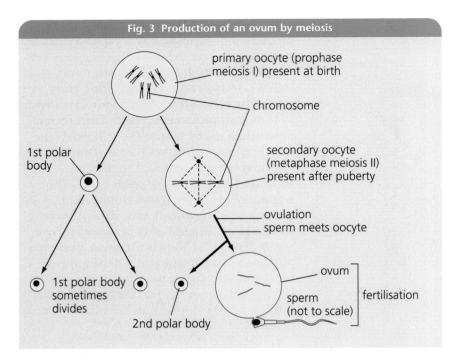

Fig. 3 Production of an ovum by meiosis

primary oocyte (prophase meiosis I) present at birth

chromosome

1st polar body

secondary oocyte (metaphase meiosis II) present after puberty

ovulation
sperm meets oocyte

1st polar body sometimes divides

ovum

2nd polar body

sperm (not to scale)

fertilisation

Ovulation

The mature follicle is 10–15 mm in diameter and is known as a **Graafian follicle**. It moves towards the surface of the ovary where it bursts and releases the secondary oocyte; this is called **ovulation**. Only one follicle usually reaches this stage during each cycle, the rest degenerate.

The oocyte is surrounded by the **zona pellucida** (jelly coat). Many granulosa cells from the follicle remain surrounding the zona pellucida after ovulation. These extra cells may help the cilia to move the oocyte into the funnel and down the oviduct.

> **2** A woman reached puberty at age 13. She later had three children and reached the menopause at age 45. After each child, her menstrual cycle stopped for three months while the baby breast-fed.
>
> **a** Calculate how many secondary oocytes were ovulated by this woman in her lifetime. State any assumptions you make.
>
> **b** State the longest time, in years, between the formation of a primary oocyte and its development into a secondary oocyte in this woman.

Hormonal control of the menstrual cycle

The release of an oocyte by the ovary is linked to the repair of the endometrium. This means that, if fertilisation occurs, the uterus

is ready to receive the embryo. These changes in the ovary and uterus are coordinated by the interaction of four hormones:

- two protein hormones, **follicle stimulating hormone (FSH)** and **luteinising hormone (LH)** made by the anterior lobe of the **pituitary gland**;

- two steroid hormones, **oestrogen** and **progesterone** made by the ovaries.

The pituitary gland is attached by a stalk to the **hypothalamus**. The hypothalamus contains specialised neuro-secretory cells, which secrete hormones rather than neurotransmitters. The hormones travel through small capillaries from the hypothalamus to the nearby pituitary gland, to control the pituitary's secretion of FSH, LH and other hormones. The gonadotrophic hormones, FSH and LH, travel in the bloodstream to the ovaries (female gonads) where they stimulate production of oestrogen and progesterone. These two ovarian hormones in turn have **feedback effects** on the hypothalamus and pituitary gland.

The start of the menstrual cycle is signalled by the beginning of menstruation on day 1 (Fig. 4). The lining of the endometrium passes out via the cervix and vagina over 4–5 days. The inner layer of the endometrium remains attached to the myometrium allowing a new endometrium to grow by mitosis later in the cycle.

Follicle development in the ovaries is triggered by the release of FSH from the pituitary gland in the first week of the cycle. As the follicle cells begin to divide, they develop a better blood supply into which they release oestrogen. The oestrogen has two effects:

- it binds to specific **receptor proteins** in the surface cells of the uterus, causing the repair and thickening of the endometrium;

- it has a feedback effect on the secretion of the pituitary hormones FSH and LH.

During days 5–10, oestrogen levels rise *slowly*. Oestrogen inhibits the release of FSH and LH, an example of **negative feedback**. FSH levels decline and LH levels remain low. During days 10–14 there is a *rapid* rise in oestrogen levels. This has a **positive feedback** effect on FSH and LH. There is a surge in FSH and LH levels around days 12–14. These surges cause the primary oocyte to develop into a

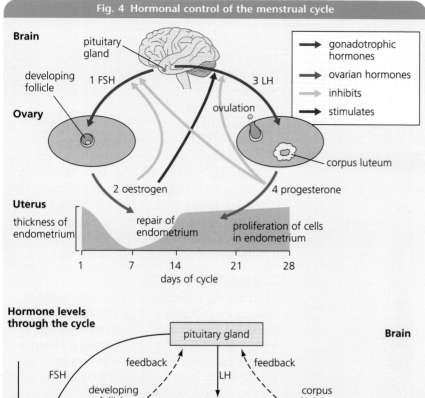

Fig. 4 Hormonal control of the menstrual cycle

secondary oocyte (plus polar body) and then trigger ovulation at day 14. LH also causes the transformation of the empty follicle cells into the **corpus luteum** (yellow body).

The corpus luteum is a mass of cells whose role is to produce progesterone and oestrogen in the second half of the cycle. From about day 15, progesterone levels start to rise, reaching a peak at about day 21. Progesterone has two effects:

● like oestrogen in the first half of the cycle, it binds to specific receptor proteins in cells on the surface of the endometrium; the endometrium cells respond by becoming glandular and secreting **glycoproteins** that make the endometrium highly receptive to an embryo on days 20–21;

● it has a feedback effect on FSH and LH

– at *low* concentrations, progesterone has a negative feedback effect on FSH, so preventing further follicles from developing during the second half of the cycle

– at *high* concentrations, progesterone has a negative feedback effect on LH, which (since LH stimulates the corpus luteum to make progesterone) effectively turns off its own production – the resultant fall in progesterone from day 21 results in constriction of blood vessels in the uterus, and the start of menstruation at the end of the 28-day cycle.

3a Outline the effects of progesterone on the endometrium and the ovary.

b State two negative feedback effects of progesterone on the pituitary gland.

KEY FACTS

■ The menstrual cycle involves the regular building up and shedding of the endometrium (uterus lining).

■ The ovaries release a secondary oocyte every 28 days. The secondary oocyte is arrested at metaphase II of meiosis.

■ The oocyte develops inside a group of cells called a follicle. Several follicles develop in each ovary every cycle, but usually only one matures into a Graafian follicle and ovulates.

■ The menstrual cycle co-ordinates the release of an oocyte with the preparation of the uterus wall to receive an embryo.

■ The menstrual cycle is controlled by the interaction of two pituitary hormones, FSH and LH, and two ovarian hormones, progesterone and oestrogen. These four hormones interact through positive and negative feedback effects.

■ The effects of these four hormones depends on both their relative concentrations and their rate of change of concentration, in the bloodstream.

Conception

During intercourse (or coitus), the penis ejaculates 3–5 cm³ of **semen** at the base of the cervix. Semen contains 20–100 million **spermatozoa** (sperms) per cm³, which are produced by meiosis in the testes, plus **seminal fluid** made by the **prostate gland, seminal**

vesicle and other **accessory glands**. Seminal fluid has several important roles:

- it is slightly alkaline and helps to neutralise the acidity of the vagina;
- it contains fructose, a sugar that provides energy for the sperms;
- it provides a medium in which sperms can swim, and causes them to become motile;
- it stimulates small uterine contractions that help to move sperms towards the oviduct;
- it coats the sperm heads with a protein that protects against the acidity and immune response of the vagina.

Capacitation

Sperm head membranes have a protective coating of proteins and glycoproteins provided by secretions in the epididymis and by the seminal fluid. The coating protects sperms from the hostile environment of the reproductive tract, and enables them to survive longer as they swim towards the oviduct. However, the coating must be removed by enzymes in the uterus before sperms have the capacity to fertilise an egg. This is **capacitation** and it takes 6–7 hours.

The acrosome reaction

Fertilisation usually occurs in the upper part of the oviduct. There is no clear evidence that human sperms follow a chemical signal to the oocyte (chemotaxis, Chap. 1) as occurs in some other species. The vast number of sperms and the relatively large size of the oocyte ensure that some sperms reach their target by chance.

Contact between a sperm and the granulosa cells around the oocyte triggers the **acrosome reaction** – the release of enzymes stored in the **acrosome** in the sperm head (Fig. 5).

One of the enzymes released is hyaluronidase, which digests the matrix of hyaluronic acid surrounding the granulosa cells. After passing between the granulosa cells, the sperm meets the zona pellucida. The zona pellucida is composed of glycoproteins, and one of these, ZP3, is a sperm receptor. The inner acrosomal membrane, which is exposed during the acrosome reaction, contains a receptor protein that binds to ZP3. This forms a neat slit through the zona pellucida allowing entry of the sperm, which lashes its tail to push through into the **perivitelline space** between the oocyte and

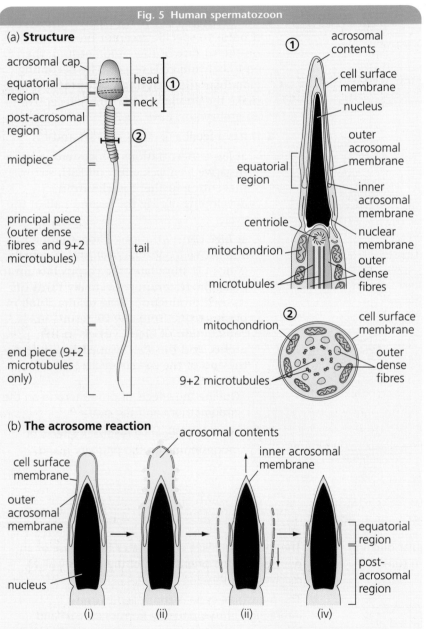

Fig. 5 Human spermatozoon

(a) Structure

acrosomal cap
equatorial region
post-acrosomal region
midpiece
principal piece (outer dense fibres and 9+2 microtubules)
end piece (9+2 microtubules only)

head ①
neck

②

tail

① acrosomal contents
cell surface membrane
nucleus
outer acrosomal membrane
equatorial region
inner acrosomal membrane
centriole
nuclear membrane
mitochondrion
outer dense fibres
microtubules

② mitochondrion
cell surface membrane
outer dense fibres
9+2 microtubules

(b) The acrosome reaction

cell surface membrane
outer acrosomal membrane
nucleus

acrosomal contents
inner acrosomal membrane
equatorial region
post-acrosomal region

(i)   (ii)   (ii)   (iv)

(i) Before the acrosome reaction. (ii) Changes in the permeability of the membranes to Ca²⁺ cause the cell surface membrane to 'point fuse' with the outer acrosomal membrane. This allows the acrosomal contents to diffuse out, releasing enzymes to help sperm penetrate the outer coats of the oocyte. (iii) The fused membranes are sloughed off, leaving the inner acrosomal membrane exposed. (iv) Physiological changes occur in the equatorial region of the sperm head, rendering it capable of fusion with the oocyte surface membrane.

Source: adapted from Baggott, *Human Reproduction*, Cambridge University Press, 1997

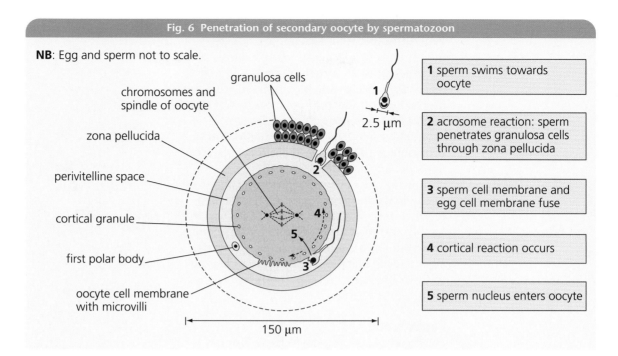

Fig. 6 Penetration of secondary oocyte by spermatozoon

NB: Egg and sperm not to scale.

granulosa cells

chromosomes and spindle of oocyte

zona pellucida

perivitelline space

cortical granule

first polar body

oocyte cell membrane with microvilli

2.5 µm

150 µm

1 sperm swims towards oocyte

2 acrosome reaction: sperm penetrates granulosa cells through zona pellucida

3 sperm cell membrane and egg cell membrane fuse

4 cortical reaction occurs

5 sperm nucleus enters oocyte

the zona pellucida. The posterior part of the sperm head fuses with the oocyte membrane, which is covered with thousands of **microvilli** that close around the sperm head. The tail of the sperm continues to move, causing the oocyte to rotate inside the zona pellucida until all the tail is pulled into the perivitelline space (Fig. 6).

The cortical reaction

It is essential that only one sperm fertilises each oocyte. If two or more sperms enter (**polyspermy**), normal development does not occur and the embryo dies after three or four cell divisions. Polyspermy is prevented by the **cortical reaction**, in which the zona pellucida forms a barrier to further sperm entry:

- fusion of sperm and oocyte causes an increase in calcium ion concentration in the oocyte cytoplasm;
- increased calcium ion concentration causes exocytosis of oocyte lysosomes, called **cortical granules**, into the perivitelline space;
- enzymes from the cortical granules cause the zona pellucida to thicken (forming the **fertilisation membrane**) and also destroy ZP3, so no more sperm can bind to the zona pellucida.

4a Explain how the enzymes in the acrosome are released at the correct time.

b Explain how polyspermy is prevented.

APPLICATION

Male infertility

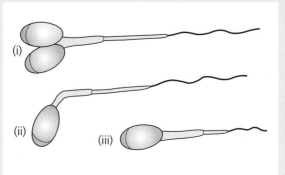

(i)

(ii)

(iii)

Men can be infertile for several reasons including:

- a sperm count below 10 million sperm per cm³;
- large numbers of abnormal sperm as shown in the diagram;
- infection of the prostate gland;
- low testosterone (male sex hormone) levels from the testis.

1 Suggest why each of these factors causes male infertility.

Formation of the zygote

When the sperm and secondary oocyte membranes fuse, the process of meiosis in the oocyte – arrested at metaphase II – is reactivated. Meiosis II is completed with the production of a second polar body and the haploid **female pronucleus**. The female pronucleus will fuse with the haploid **male pronucleus** to form the diploid zygote. The male pronucleus forms from the chromosomes in the sperm head. The fusion of the male and female pronuclei is called **fertilisation**. Within 24 hours of fusion, the female and male pronuclei membranes break down, their chromosomes mingle, and the zygote immediately begins its **first cleavage** to form the **embryo** (Fig. 7). The process of fertilisation is now complete.

Implantation

During cleavage, the developing embryo is moved by cilia down the oviduct. For the first five days, the embryo remains surrounded by the zona pellucida, which keeps the embryo cells together and stops the embryo from

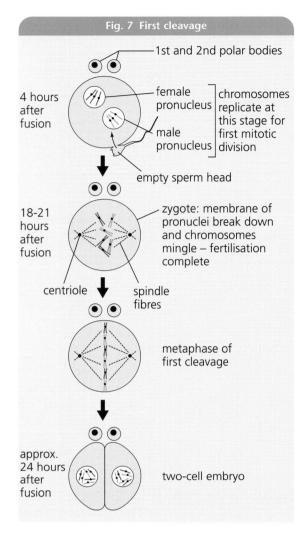

Fig. 7 First cleavage

1st and 2nd polar bodies

4 hours after fusion

female pronucleus — chromosomes replicate at this stage for first mitotic division

male pronucleus

empty sperm head

18-21 hours after fusion — zygote: membrane of pronuclei break down and chromosomes mingle – fertilisation complete

centriole   spindle fibres

metaphase of first cleavage

approx. 24 hours after fusion — two-cell embryo

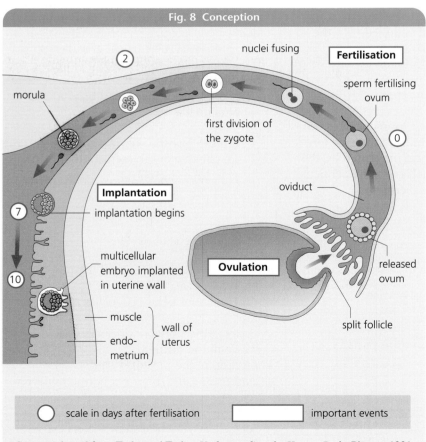

Fig. 8 Conception

②

morula

nuclei fusing

Fertilisation

sperm fertilising ovum

first division of the zygote

⓪

Implantation

implantation begins

⑦

oviduct

multicellular embryo implanted in uterine wall

Ovulation

released ovum

⑩

muscle

endo-metrium

wall of uterus

split follicle

○ scale in days after fertilisation   ▭ important events

Source: adapted from Tudor and Tudor, *Understanding the Human Body*, Pitman, 1981

implanting in the oviduct wall. By this time, the embryo is a ball of more than 100 cells called a **morula** and is slightly smaller than the original zygote. The embryo now develops a central fluid-filled space and is called a **blastocyst**. About seven days after fertilisation, the blastocyst implants in the uterus wall (Fig. 8). Outer cells of the blastocyst, called **trophoblast cells**, invade the uterus wall, until the embryo becomes completely buried in the endometrium. Once successful implantation is complete, **conception** has occurred: the woman is pregnant. The trophoblast cells develop into a membrane called the chorion, which plays a major role in the exchange of materials between mother and developing embryo. The placenta arises from a combination of maternal and fetal tissues. The presence of two placentas indicates twins, though identical twins sometimes share a single placenta.

APPLICATION

Birth weight effects

This figure shows rates of coronary heart disease in a total of 517 adults aged 45+ years in South India, plotted against their birth weight.

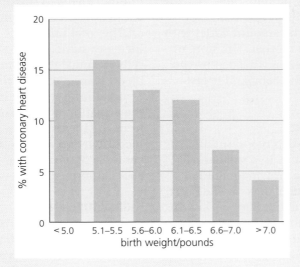

1 a Outline the way in which these data could have been obtained.

b Describe the relationship between birth size and adult heart disease.

c Suggest a possible explanation for this relationship.

Small babies are often less healthy than babies of normal size, and are more likely to die around birth. It has been found that larger mothers tend to have larger babies, but that the size of the father has little effect on birth size.

2 Suggest why it is the mother's size, not the father's size, which mostly influences the baby's birth size.

KEY FACTS

■ The placenta forms a physical link between mother and fetus in which the two circulations pass close to each other over a large surface area, but do not mix.

■ Exchange of materials occurs across thousands of chorionic villi by diffusion, osmosis, active transport and pinocytosis.

■ A layer called the trophoblast (which originates from the outer cells of the blastocyst) covers the exchange surface and helps to control the passage of substances from mother to fetus.

3.5 Physiological changes in the mother

A positive result – pregnancy is generally a very happy time for both parents.

Pregnancy is very demanding both physically and emotionally, and involves dramatic physiological changes to the mother. The mother gains weight as her uterus, fetus, placenta and breasts grow. The abdominal muscles are stretched and internal organs are displaced by the growing fetus. Movement becomes more difficult, morning sickness may occur as a result of hormonal changes, and the mother must prepare herself psychologically for the stresses of labour and motherhood. The father also needs to prepare himself mentally, and the relationship between the two parents must adapt appropriately, especially if it is their first child.

Table 2 Typical mass changes in pregnancy	
Source	Mass increase /kg
uterus	1.0
fat (variable)	4.0
amniotic fluid	0.5
tissue fluid	2.5
placenta	0.5
fetus	3.5
total	12.0

Body mass

The average gain in body mass during pregancy is about 12 kg (Table 2).

About 1 kg of the mass increase is protein, half of which is found in the fetus and placenta, and half in the myometrium, breast glandular tissue, plasma protein and haemoglobin. Total body fat also increases during pregnancy, but the amount varies with total weight gain of the mother. The store of fat helps to supply the energy needed by the mother for the care of the baby and breast-feeding.

Cardiovascular changes

The placenta requires a rich blood supply: an inadequate placenta results in a smaller baby or even miscarriage. The growing maternal tissues also make extra demands on the heart. Cardiac output rises by about 30% during the first 10 weeks of pregnancy and by 40% by weeks 20–24, then it stabilises.

The higher output is due partly to a 12% increase in the heart's size, which allows the stroke volume to increase: the consequent increase in cardiac output can reach 1.5 dm³ per minute over the non-pregnant level.

Blood pressure declines slightly during pregnancy, particularly diastolic pressure (blood pressure when the ventricles relax) which is reduced by 5–10 mm Hg. However, blood plasma volume rises by up to 50% to provide extra blood flow through the placenta, kidneys and other organs. About 10% of the mother's blood flows through the placenta for each circuit round the body.

Gas exchange

Improved heart function increases the supply of oxygen and the removal of carbon dioxide via the placenta, as well as speeding up the provision of nutrients and removal of urea. Gas exchange is further increased by improvements to the mother's lung function and red blood cell numbers.

The total number of red blood cells increases sufficiently to allow the number cells per mm³ to remain roughly constant as blood plasma volume rises.

Total lung capacity is slightly reduced by the elevation of the diaphragm by pressure from the uterus. However, tidal volumes increase by up to 50%, allowing the mother's basal metabolic rate to increase by 15–20% above non-pregnant levels.

Kidney function

During pregnancy, blood flow through the kidneys rises by 25–50%, and glomerular filtration rate increases accordingly. The kidneys grow in length by about 1 cm.

The volume of urine passed by the mother per day is not significantly increased, showing that the kidneys are producing more concentrated urine during pregnancy. This removes the extra urea produced by amino acid metabolism in the livers of mother and fetus.

The fetus' kidneys start to function during the 10th week of development. From this time, the fetus loses some urea by urination into the amniotic fluid, which will be lost at birth.

KEY FACTS

- Pregnancy requires major physiological and psychological changes in the mother.
- Maternal body mass increases by about 12 kg through the growth of the uterus, placenta, fetus, blood, breasts and other organs.
- The mother's heart increases its capacity allowing for improved circulation through the placenta, lungs and other organs.
- Extra maternal red blood cells are synthesised and lung function increases, enhancing gas exchange with the fetus.
- Kidney size and filtration rate increase in the mother during pregnancy allowing the removal of the extra urea produced by the fetus and mother during pregnancy.
- During pregnancy, the mother's body is also preparing for giving birth and caring for the child. Pregnancy and child bearing are a time when a mother needs a lot of emotional and physical support. The father and other members of the family can be of crucial importance here; they can provide a loving and stimulating environment for the mother and newborn child, encouraging the child to grow into a well-adjusted adult. A positive pregnancy test is just the first step along the lifelong road of a parent.

1

a On the axes provided, sketch a graph to show the changes in the concentration of LH (luteinising hormone) in the blood during a **typical** menstrual cycle. (2)

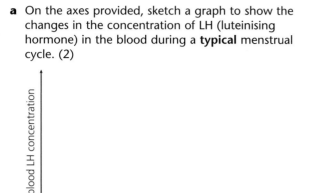

b In an experiment to investigate the control of LH release from the pituitary gland, a female mammal was injected with a large dose of oestrogen. The concentrations of LH in the blood were measured at 12-hourly intervals before and after this injection. The figure shows the blood LH concentrations measured during the investigation.

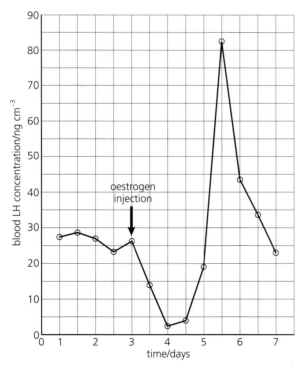

i) Describe the changes in blood LH concentration in response to the oestrogen injection. (3)

ii) Explain the role of oestrogen in the control of LH release in this investigation. (2)

iii) Relate your answer in b ii) to the pattern of LH release in the typical menstrual cycle. (2)

OCR/GDR 4805 March 1996 Q2

2

a State three factors which will affect the rate of exchange of substances between mother and fetus across the placenta. In each case, explain the effect of each factor. (3)

Below are diagrammatic representations of placental structure in humans and pigs. The diagram shows details of the structure in the region which exchanges take place between fetal and maternal blood.

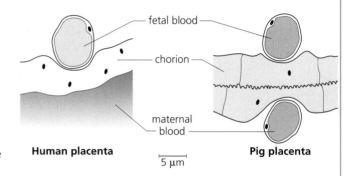

Human placenta   5 μm   **Pig placenta**

b i) State two important structural differences between the two placentas. (2)

ii) Suggest how these differences might affect exchanges across the placenta in each animal. (2)

The table below shows the oxygen content and percentage saturation of maternal and fetal blood in four different blood vessels.

	Maternal blood vessels		Fetal blood vessels	
	uterine artery	uterine vein	umbilical artery	umbilical vein
% saturation of blood with oxygen	95	70	25	65
oxygen content/cm^3 per 100 cm^3 blood	14	10	5	13

c Explain the differences in the oxygen content of the blood in each of the four vessels. (4)

OCR/GDR 4805 November 1996 Q2

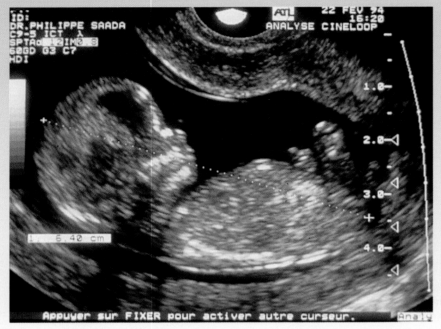

Most people would like to have a choice about when to start a family. Some couples prefer to delay having children until they have spent some time establishing their relationship, buying a house or developing a career.

The WHO estimates that there are over 100 million acts of sexual intercourse each day resulting in almost 1 million pregnancies of which about 50% are unplanned and 25% are unwanted. At the same time, there are large numbers of people who wish to have children but cannot do so. There is clearly an urgent need for further research and investment in reproductive technology, but we must remain sensitive to public opinion and the ethical issues involved. Control over this aspect of our lives is one of the greatest challenges for the future.

If we wish to plan our families and overcome any problems of infertility, we need to be able to intervene in the reproductive process. Successful **contraception** and **fertility treatment** are based on an understanding of human reproduction and on improvements in technology to make methods effective and safe.

4.1 Birth control and contraception

For most couples, **birth control** is the main concern. It includes:

- the many ways of preventing conception (contraception);
- methods of removing a fetus after conception (e.g. abortion).

Hormonal methods of contraception

Women can take female hormones in tablet form to control their fertility. This is called oral contraception and is the most effective and reliable type of birth control other than sterilisation by surgery. There are several varieties of oral contraceptive for women. Most contain a relatively high level of oestrogen. This inhibits the production of FSH by the pituitary gland. Without FSH, no follicle develops and ovulation does not occur, so conception is not possible. Some contraceptive pills, called combined pills, also contain artificial progesterone called progestin.

Progesterone inhibits the production of LH, thereby further reducing the likelihood of ovulation, but the role of progesterone in hormonal contraception is more complex than that of oestrogen. Contraceptive pills are taken once a day for 21 days of the 28-day cycle. During the seven days when the pill is not taken, the levels of oestrogen and progesterone fall, and the endometrium breaks down. This produces a lighter than usual menstruation.

The minipill contains only progestin, which is chemically similar to progesterone and can mimic the action of progesterone. It also works by thickening the cervical mucus, preventing sperm entry. The morning-after pill (emergency contraception) contains large doses of progestin and oestrogen, or progestin alone. It is now available without prescription, and is about 75% effective in preventing implantation if taken early enough.

Hormonal implants can be inserted under the skin of the upper arm to give contraceptive protection to women. Six matchstick-sized rubber rods are implanted in a fan pattern and steadily release progestin into the bloodstream for 3–5 years. This kit contains everything needed for the procedure.

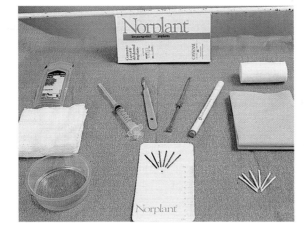

injection or implant to counteract the reduced output from the testes, which would otherwise cause a loss of sex drive and secondary sexual characteristics. Preventing sperm production without side-effects is not easy, and a male pill is still not commercially available.

Other approaches to male contraception include targeting the acrosome reaction or paralysing the vas deferens muscles. In addition, there are genes involved in controlling sperm maturation which it may be possible to inactivate.

2a Suggest three essential features of a male contraceptive pill if it is to be widely accepted and used.

b Give three disadvantages of using a male contraceptive pill rather than a condom.

1a Explain why progesterone in oral contraceptives prevents ovulation.

b Progestin is similar, but not chemically identical to progesterone. Suggest why it is possible for progestin to work as a contraceptive.

The male pill

Human trials are underway to test a **male pill** that contains progestin and a small amount of testosterone. The combined effects of these hormones is to inhibit – by negative feedback – gonadotrophin secretion by the pituitary gland. This results in a reduction in testis activity. Some extra testosterone is supplied by

Barriers, sterilisation and IUDs

These contraceptive methods are collectively known as non-hormonal methods (Fig. 1). Barriers can be physical (condom and diaphragm) or chemical (spermicide creams and pessary tablets that are placed in the vagina before intercourse to kill or immobilise sperm; usually used in addition to a barrier method). These methods all aim to prevent the sperm and oocytes from meeting.

Sterilisation is considered permanent, because it is very difficult to reverse. Sterilisation for women usually involves putting a clip on the oviducts so that there is no passage from the ovary to the uterus. This is called **tubal ligation**. For men, sterilisation is by removing or tying a small section of the vas deferens. This is called a **vasectomy**.

The intrauterine device (IUD) or 'coil' is a modern development of the ancient discovery that the presence of a small object in the uterus can prevent conception. A modern IUD is kept permanently in the uterus. There are two main types – those that contain copper and those that contain progesterone. IUDs prevent implantation or immobilise sperm, and non-hormonal ones are more effective when they contain copper. Exactly how IUDs work is not known.

Natural methods

Natural methods of birth control are based on the practice of avoiding intercourse during the most fertile part of the woman's monthly cycle – immediately before ovulation. A woman's body temperature increases slightly at the time

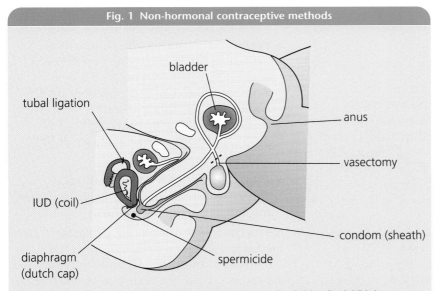

Fig. 1 Non-hormonal contraceptive methods

Source: adapted from Gadd, *Individuals and Population*, Cambridge Social Biology Topics, Cambridge University Press, 1983

of ovulation, so it is possible to identify when ovulation has occurred by keeping accurate body temperature records (Fig. 2).

3 Suggest three reasons why using body temperature measurements is an unreliable method of birth control.

Lactation, or breast-feeding, can have a contraceptive effect because it causes changes in the levels of the hormones that control ovulation and thereby reduces the chances of pregnancy. Lengthy lactation is common in some communities; for example, some hunter–gatherer mothers continue to suckle their babies for up to two years after birth, during which time they do not become pregnant.

Comparison of methods

The effectiveness of various birth control methods can be compared using the number of pregnancies occurring in a year for every 100 women using the method (Table 1). The most suitable method for any couple depends on a number of factors including the following.

Table 1 Effectiveness of some birth control methods	
Method	**Pregnancies per year per 100 women**
no method	50
natural methods	17–22
spermicide alone	20–30
diaphragm with spermicide	2–9
condom	2–7
IUD	2–4
combined pill	0.1
sterilisation	less than 0.05

- ***Coping with an unplanned pregnancy*** A young couple who do not yet have the income or accommodation to support a baby need the most reliable form of contraception, such as oral contraception or a hormone implant.

- ***Reversibility*** If a couple have all the children they want, then sterilisation (vasectomy or tubal ligation) may be advisable. Care must be taken over long-term use of hormone methods as in some cases the menstrual cycle does not easily restart.

- ***Medical condition*** A woman who has a family history of **thrombosis** might choose not to use hormone-based contraceptives as these can potentially increase the risk of this condition. An IUD should be avoided if a woman has a history of pelvic inflammatory disease or other infections.

- ***Duration*** If protection is needed without worry for a long time, then an IUD, hormone implant or sterilisation may be suitable.

- ***Protection from sexually transmitted disease (STD)*** If contracting an STD is a concern, a barrier method, such as a condom, should be used, possibly in addition to another method.

- ***Cost and availability*** This is a very significant factor in developing countries and poor communities.

- ***Ethics and religion*** Some methods of contraception are not acceptable to people for moral or religious reasons.

No method of contraception is ideal and new methods are urgently needed.

There are an estimated 50–60 million abortions carried out each year worldwide, indicating the scale of failure of current contraceptive control.

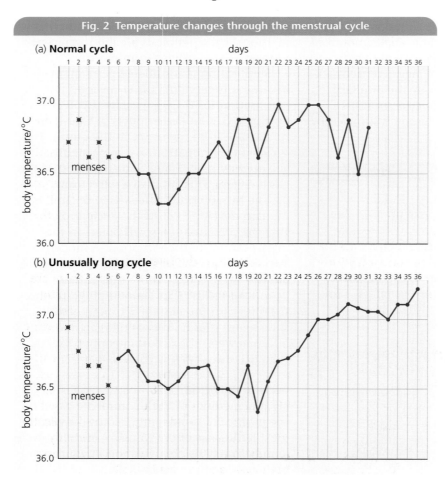

Fig. 2 Temperature changes through the menstrual cycle

(a) **Normal cycle**

(b) **Unusually long cycle**

Other contraceptive ideas

In about 5% of men, antigens on the surface of the sperm trigger an immune response in the man's body; these men are infertile. This suggests the possibility of using a vaccine as a means of contraception. The vaccine would be targeted at the sperm head's receptor proteins or at the receptor molecules (ZP3) on the zona pellucida, which are involved in the acrosome reaction (Chap. 3).

Generations of women in Sri Lanka and India have eaten the papaya fruit to try to prevent pregnancy. When they wished to become pregnant, they simply stopped eating the fruit. It is claimed that women can even terminate pregnancy by eating the fruit. Papaya contains papain, an enzyme that attacks progesterone, so it is possible that the method works.

1 a Describe what happens when a sperm meets ZP3 molecules.

   b Suggest how a vaccine could inactivate this stage of the acrosome reaction.

2 a Explain how papaya could act as a contraceptive.

   b Suggest a reason why ingestion of a hormone-digesting enzyme might not work as a means of contraception.

- Contraception can be used to control the time and number of children born.

- Contraception can involve natural methods, hormones, barriers, IUDs or sterilisation. New methods are being researched and are urgently needed.

- Every method of contraception has advantages and disadvantages. The most suitable method for any couple depends on medical factors, age, existing family size, and financial and social circumstances.

4.2   Infertility

When a couple decide to start a family, it is quite common for the woman not to conceive for several months or even a few years. In the UK, about one in six couples now seeks help for childlessness. Female infections such as pelvic inflammatory disease or chlamydia are common causes of infertility. Such infections may go unnoticed, but can lead to blocked oviducts and other problems. In males, a low sperm count may be responsible. There are some concerns about a possible decline in

average male sperm count and sperm quality over the past 50 years, though the evidence is still inconclusive. Often, the cause of a couple's infertility cannot be found, but they may still be helped (Table 2).

Assisted reproductive technology

Fertility treatment is big business. There are now private clinics specialising in helping childless couples all over the world, and treatments are also available from some state health services. One of the commonest treatments is *in vitro* fertilisation (IVF), which hit the headlines with the birth of the first 'test-tube baby' in 1978. Today, more than 1% of children born in the UK are test-tube babies. *In vitro* means 'in glass', and indicates that fertilisation occurs outside the body, usually in a glass dish. IVF is used to treat both male and female infertility.

Table 2 Common causes of infertility	
Male	**Female**
low sperm count	oviduct blocked
poor sperm motility	ova not viable
deformed sperm	hormone imbalance
erectile problems	poor implantation

The petri dish being held by the gloved hand contains fluid with oocytes. The oocytes will be selected and removed under the microcscope for later fertilisation.

Hormone treatment

A couple who seek IVF treatment are interviewed and assessed before medical tests begin. IVF can use the oocytes a woman produces in her natural cycle, but a higher success rate is achieved if she is given hormone-based drugs to induce super-ovulation – the production of several ripe follicles in one cycle. The combination of drugs used depends on the individual, and may include clomiphene citrate, artificial pituitary releasing hormones, FSH and LH (Chap. 3). The normal menstrual cycle is suppressed, then ovulation is induced at a time controlled by the doctor.

Progress is checked from about a week later. The woman's hormone levels are monitored and her ovaries are checked by **ultrasound**. When the follicles are almost

ripe, hCG is given by injection, which helps the oocytes to mature.

Collecting the gametes

An ultrasound probe is used to guide a hollow needle towards the mature follicles in the ovary, usually entering through the vagina wall. The oocytes are removed from the follicles by suction.

A semen sample is now provided by the man. He should not have ejaculated for 3–7 days, to ensure a good quality sample. The semen is treated to select the most vigorous sperm, and a sample containing about 100 000 sperm is added to each oocyte. The sperms and eggs are cultured together in a petri dish in a suitable fluid medium overnight. The eggs are then inspected by a microscopist who looks for the pronuclei of the zygote or 2–4 celled embryos.

There are several micro-manipulation techniques for increasing the chances of fertilisation (Fig. 3). When a man has severe fertility problems, intracytoplasmic sperm injection (ICSI) is often used; a single sperm is injected into the oocyte. Despite some anxiety over potential abnormalities arising from the use of defective sperm for fertilisation, about 20 000 healthy babies have been born using ICSI over the last decade.

4 Suggest how the most vigorous sperm could be selected from a semen sample for IVF treatment.

Replacing the embryos

The healthiest embryos are selected and placed into the upper uterus using a catheter. With luck, one or more embryos will implant and develop.

In the UK, the Human Fertilisation and Embryological Authority (HFEA) sets out guidelines and rules for fertility treatment. A maximum of three embryos is permitted at one time, to prevent the dangers of multiple births. Many clinics use just two, as the success rate is almost as good as with three (Table 3).

Embryos that are not used may be frozen and stored for up to 5 years, and semen for 10 years. About 75% of frozen embryos survive the thawing process. Some clinics are now developing methods of freezing and thawing unfertilised eggs, but this is more difficult.

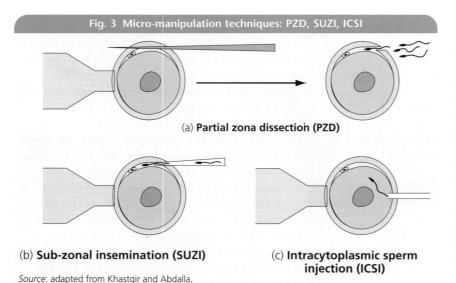

Fig. 3 Micro-manipulation techniques: PZD, SUZI, ICSI

(a) **Partial zona dissection (PZD)**

(b) **Sub-zonal insemination (SUZI)**

(c) **Intracytoplasmic sperm injection (ICSI)**

Source: adapted from Khastgir and Abdalla, 'Assisted conception', *Clinical Review of Gynaecology*, 1994

Table 3 IVF (including ICSI) pregnancy rates in relation to the number of fresh embryos transferred			
Number of embryos transferred	Number of embryo transfers	Number of pregnancies	Pregnancy rate per embryo transfer
1	56	6	10.7
2	263	56	21.3
3	290	77	25.7

Source: Bourn Hall Clinic website, July 2000

Success rates

IVF is particularly helpful for women with blocked oviducts. The overall success rate of IVF has steadily improved, from about 10% of live births per fertility cycle in the 1980s to over 20% today. The success is much lower with older women (Table 4). The cost of each treatment is £2000–3000.

Despite recent advances, IVF remains a costly process with a limited success rate. Implantation can be enhanced by making a small hole in the zona pellucida, assisting the embryo to 'hatch' and implant. Further improvements are likely. One promising approach is blastocyst transfer. This technique uses 5–6-day-old embryos in the blastocyst stage rather than the 3-day-old embryos usually employed. Implantation naturally occurs at the 5–6 day stage, so the technique may improve implantation rate and also allow better embryo selection.

Embryo selection can now involve pre-implantation genetic diagnosis (PGD) whereby one or two cells are removed from an 8-cell embryo for genetic analysis. The remaining part of the embryo will grow normally and the removed cells can be screened for factors such as:

- chromosome abnormalities;
- gender;
- genetic markers, such as for mutations in the breast cancer gene;
- a compatible tissue donor.

We are still a long way from being able to select embryos for stature or intelligence, and this is likely to remain expensive and controversial. Selection for gender to avoid sex-linked diseases, and for compatible donors for existing sick children have already occurred, and are the subject of much debate.

Fertility drugs alone

Infertility may be treated in more simple ways. For example, women who fail to ovulate regularly, can be given clomiphene to block oestrogen inhibition by FSH, giving a greater chance of egg production.

Ethics

Advances in reproductive technology introduce all kinds of ethical issues. The welfare and rights of the parents and their children must be carefully considered. The HFEA requires clinics to take account of the welfare of the potential child as well as that of other members of the family. New Human Rights Legislation in Europe may lead to legal challenges to existing rules. These new laws include the principle that no-one who wants a child should be denied the right to have one, regardless of sexual orientation, age or other factors.

The fate of stored embryos and semen is also problematical. Sometimes, the owners of samples have died or cannot be traced. Should these or other samples be available for donation or research? Rules on embryo

Table 4 IVF (including ICSI) clinical pregnancy rates in relation to female patient's age							
Age	Cycles	Egg collections	Embryo transfers	No. of pregnancies	Pregnancy rate per cycle/%	Pregnancy rate per egg collection/%	Pregnancy rate per transfer/%
25	6	5	3	1	16.6	20	33.3
25–29	97	91	68	19	19.6	20.9	27.9
30–34	332	306	252	75	22.6	24.5	29.8
35–39	283	264	215	49	17.3	18.6	22.8
40+	91	86	71	7	7.7	8.1	9.8

Source: Bourn Hall Clinic website, July 2000

research have recently been relaxed in the USA and in the UK, allowing the cloning of **stem cells**. Stem cells can potentially grow into any type of tissue or organ that could then be used for transplantation into adults with problems such as Parkinson's disease or diabetes. Stem cells might even be a source of eggs or sperm in future.

IVF and pregnancy rates

This table shows pregnancy rates achieved at a particular clinic by IVF (including ICSI) in relation to the cause of infertility.

1 What is the commonest reason for fertility treatment at this clinic?

2 Suggest why:

a the number of egg collections was lower than the number of cycles treated;

b the number of embryo transfers was lower than the number of egg collections;

c the number of pregnancies was lower than the number of embryo transfers.

3 Which stage of the process is the least successful? Use figures to support your answer.

4 Explain why tubal damage commonly causes infertility.

5 Suggest why it is inadvisable to transfer more than two embryos.

Diagnosis	Cycles	Egg collections	Embryo transfers	No. of pregnancies	Pregnancy rate per cycle/%	Pregnancy rate per egg collection/%	Pregnancy rate per transfer/%
unexplained	155	144	119	30	19.4	20.8	25.2
male factor	248	230	169	42	16.9	18.3	24.8
miscellaneous	149	138	111	26	17.4	18.8	23.4
male and female	58	54	46	13	22.4	24.1	28.3
tubal damage	199	185	164	40	20.1	21.6	24.4
total	809	752	609	151	18.7	20.2	24.85

Source: Bourn Hall Clinic website, July 2000

KEY FACTS

■ Infertility can result from causes such as blocked fallopian tubes or defective sperm.

■ IVF can overcome some types of male and female infertility.

■ IVF involves the removal of several oocytes from the ovaries. The oocytes are then fertilised, using a semen sample, outside the body. The most healthy of the resulting embryos are placed in the uterus, in the hope that they will implant and develop.

■ For males with defective sperm, ICSI, in which one sperm cell is injected directly into an oocyte, can be used.

■ Unused embryos and semen samples may remain after infertility treatment. Deciding about the use of these samples is one of the many ethical issues arising from assisted reproductive technology.

1 The rhythm method of contraception works by avoiding intercourse on days when the chance of conception is high. To do this, a woman has to determine the day of her menstrual cycle on which ovulation occurs.

a With reference to the graph, explain the chances of conception observed when intercourse occurs on day 11 or 12. (3)

b Suggest **two disadvantages** of the rhythm method of contraception. (2)

c **Assuming no other form of contraception is being used**, suggest **two** factors, apart from defects in sperm or egg, that will affect the chances of conception once intercourse has taken place. (2)

OCR/GDR 4805 March 1997 Q1

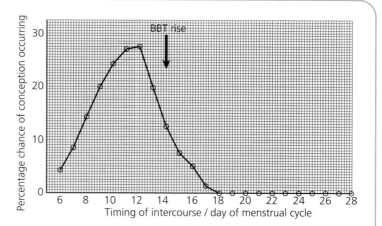

2

a Outline the main events involved in the use of *in vitro* fertilisation (IVF) to solve fertility problems. (4)

b Some men produce semen that contains spermatids instead of the mature sperm found in normal semen. This condition, in which the sperms have no tails, is called azoospermia. Men who have this condition are infertile. In the past few years, a procedure called intracytoplasmic injection (ICSI) has been used to treat this infertility. In this technique, a spermatid can be injected directly into a human egg. The babies born following this treatment have been normal, with no trace of genetic or morphological abnormality.

Some scientists have raised objections to this procedure. They claim that the procedure has not been tested sufficiently and they suggest that the use of sperm from anonymous donors would be a more suitable technique. Another concern is that the ICSI technique uses single spermatids chosen at random. This eliminates the normal competition between sperm that precedes fertilisation.

i) Explain why azoospermic men are infertile. (2)

ii) State one way in which ICSI differs from conventional IVF. (1)

c Explain what is meant by 'competition between sperm'. (2)

d State **two** ethical problems raised by the spermatid injection technique (ICSI). (2)

OCR/GDR 4805 November 1996 Q3

3

a Describe the events taking place during the acrosome reaction. (4)

b Explain how a knowledge of the acrosome reaction might be used to develop a male contraceptive pill. (2)

4

a Compare the structure of a secondary oocyte with that of a sperm cell (spermatozoon). (6)

b Describe the events that occur when a sperm meets a secondary oocyte in the oviduct, up to the time of fertilisation. (6)

c Explain how abnormalities in sperm count and sperm structure can lead to male infertility. (4)

d Outline the techniques that can be used to overcome male infertility. (4)

5 Human growth and development

Have a good look at people around you today and consider in what ways humans change as they get older. The human body grows and develops during a person's lifetime. Growth is shown by an increase in height and mass, and is most dramatic during childhood.

As humans get older, they 'age'. Biologically, **ageing** really begins as a embryo and includes all of the changes in growth and development in the lifetime of an individual. However, in humans it is ageing in adulthood that is of particular interest.

Ageing in adulthood causes some structures and body functions to deteriorate. This deterioration with time varies between organs and between individuals. This raises questions such as: What causes growth and development? Are such changes caused mainly by genes, by the environment, or a combination of the two? What causes ageing and can it be slowed down? If we could answer these questions, it might be possible to increase the length of a person's life. But can the quality of life be maintained with an increasing number of old people in the population?

This 80-year-old man has lived through many changes in China; how long will his grand-daughter live?

5.1 Patterns of human growth

A simple way of monitoring growth is to measure height or mass. By plotting size against time, the pattern of growth is shown as a **growth curve** (Fig. 1a). **Growth rate** is the change in size per unit time. Although humans grow from birth until they are adults, the growth rate is not even. Males and females show a similar pattern of growth, although the rate and timing is not the same for both sexes (Fig. 1b).

Following birth, we can divide the human life into the four phases shown in Table 1. Changes in growth rate and physical appearance occur in each phase.

Table 1 Phases of life	
Life phase	**Approximate age/years**
infancy	up to 4
childhood	4–11
adolescence	11–18
adulthood	18+

1 Study Fig. 1a. Compare the growth of males and females during infancy, childhood, adolescence, and adulthood.

2 Study Fig. 1b. Give the two ages when the growth rate is greatest in males, and in females.

Fig. 1 Mass, age and growth rates for girls and boys

(a) Growth and age

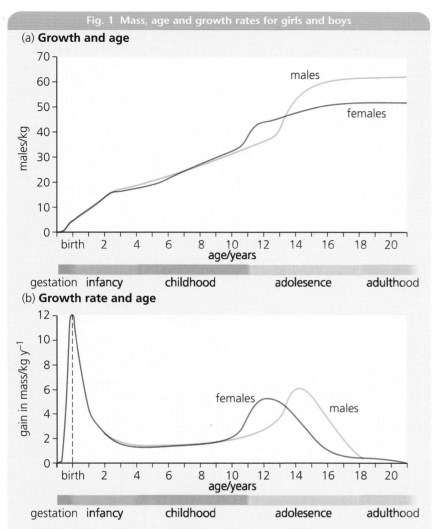

males

females

males/kg

age/years

gestation infancy childhood adolesence adulthood

(b) Growth rate and age

gain in mass/kg y⁻¹

females

males

age/years

gestation infancy childhood adolesence adulthood

Source: adapted from Gadd, *Individuals and Population*, Cambridge Social Biology Topics, Cambridge University Press, 1983

Fig. 2 Age and body proportions in boys and girls

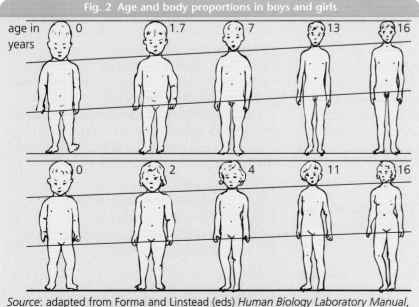

age in years

0   1.7   7   13   16

0   2   4   11   16

Source: adapted from Forma and Linstead (eds) *Human Biology Laboratory Manual*, Science Teachers Association of Western Australia, 1984

These identical twins live on their father's farm in Nepal.

There is considerable variation between individuals due to genetic and environmental factors. Studies on identical twins examine the effects of these factors.

Identical twins begin life as a single fertilised ovum, which divides to form two genetically identical individuals. Studies show that identical twins brought up in the same environment have almost identical patterns of growth and development. However, growth and development are not the same if identical twins are reared apart. We can conclude from this that the environment affects the way in which genes are expressed.

Allometric growth

Many parts of the body grow at different rates and this differential growth is known as **allometric growth**. It causes a change in body proportions between infancy and adulthood. By scaling pictures to the same size, we can see how the relative proportions of the main parts of the body change with time (Fig. 2).

3a Measure the lengths of the head, body and legs for each drawing in Fig. 2. Now measure the total length (head, body, and legs) of each figure. Calculate the percentage length of the head, body and legs for each figure. Tabulate your results.

b Describe the change in percentage length of head, body and legs during the first 16 years for boys and for girls.

The changes in body proportions are very important. The head and brain develop very quickly in the human fetus. After birth, the head and brain continue to grow but there is more rapid growth of other body parts enabling the infant to feed, move, explore,

51

and respond more independently. During adolescence, a very rapid growth spurt occurs in both sexes, accompanied by development of the reproductive organs and the secondary sexual characteristics.

As humans grow, the head becomes smaller in relation to overall size. But relative to overall body size, the human adult has a brain that is two to three times bigger than in other primates, and over seven times larger than in other mammals.

Throughout childhood and adolescence, internal organs also show allometric growth (Fig. 3). Lymphoid tissue is an important part of the **immune system** and grows most rapidly during childhood. Brain growth is most rapid in early life. The growth of the reproductive organs is most rapid during adolescence, and is accompanied by a general growth spurt. In adulthood, the body is sexually and physically mature, and growth slows down.

4a How much bigger is lymphoid tissue in an adolescent than an adult?

b Why is it important to have more lymphoid tissue during childhood?

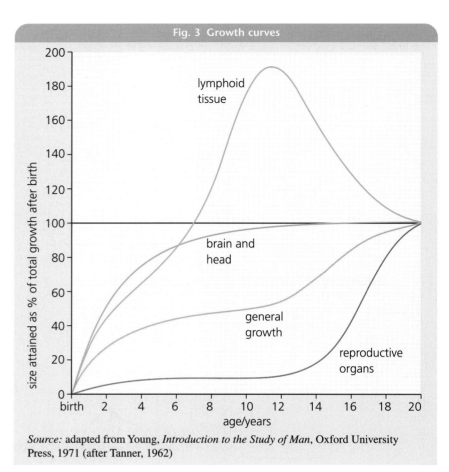

Fig. 3 Growth curves

Source: adapted from Young, *Introduction to the Study of Man*, Oxford University Press, 1971 (after Tanner, 1962)

Gestation and birth

The time from conception to birth is called the gestation period. A long gestation period allows more time for growth and development, which suggests that large babies would have better survival chances than smaller, less physically developed babies. However, the birth of a larger baby would be more difficult, with risks of injury or death to the child and its mother.

The average length of the gestation period differs in animals. We can compare the length of the gestation period as a proportion of the animal's lifespan. Compared with most animals, the gestation period of humans is relatively short. At first sight this appears to be a disadvantage, as poorly developed babies would have less chance of survival after birth.

A large brain is important in humans to develop physical and social skills, which are unique to our species. A new-born baby with large brain would be at an advantage for acquiring these skills. However, having a large brain has a disadvantage, as a larger head makes birth more difficult. Natural selection balances these two factors, the size of the baby (particularly the head) against ease of birth. Human babies are born physically immature, but with a big head compared to their body size.

5 Birth weight is believed to be determined by a combination of genes and environment. Study Fig. 4.

a Describe the relationship between birth weight and mortality.

b Describe how natural selection could affect birth weight.

c Suggest two environmental factors that could affect birth weight.

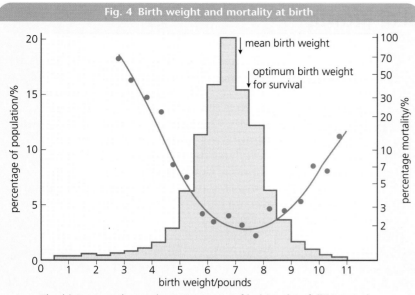

Fig. 4 Birth weight and mortality at birth

The histogram shows the percentage of babies that fall into each birth weight class. The curve plots mortality around the time of birth against birth weight.

Source: adapted from Tomkins, *Heredity and Human Diversity*, CUP, 1989

Childhood

In the **pre-pubertal period**, children have time to grow, develop, and acquire knowledge and complex skills including speech and language. The length of the pre-pubertal period as a proportion of the lifespan is longer in humans than in other animals.

However, even sexually mature offspring rely on their parents. This **extended period of dependency** is very important in the life of a human. It requires a big investment of time and energy from the parents before their offspring are physically and socially independent. The benefits are that parents can have small numbers of offspring, most of whom reach maturity with the knowledge and skills that are essential for life in a complex human society.

Birth is followed by a long juvenile phase we call childhood.

APPLICATION   Measuring growth and development

Measuring growth and development presents many difficulties because species are adapted to their natural environment, and shaped by natural selection over many generations. Large samples have to be studied to accommodate individual variation. Care must also be taken to ensure that any observations do not affect the data being recorded.

To make comparisons, measurements of physical dimensions, such as length or mass, can be made during certain developmental phases that are common between species. For example, gestation is the time between conception and birth, and the pre-pubertal phase is from birth to onset of fertility.

1 Study the table below and calculate the gestation period and pre-pubertal stage as a percentage of the lifespan, for each animal.

2 Suggest why the percentage gestation period of the human is different from the other two animals.

3 What is the significance of the percentage pre-pubertal stage in primates?

Animal	Lifespan/years	Gestation period/months	Pre-pubertal stage/years
human	75	9	14–16
chimpanzee	38	8	7
rat	3.5	0.90	0.18

5.2 Hormonal control of growth

Childhood

During childhood, the growth rate is about the same in boys and girls. The endocrine glands involved in the control of growth and metabolism during childhood include the pituitary, adrenal glands, and thyroid gland. These grow in size as the rest of the body grows, resulting in a steady growth rate.

There is an increase in the amount of body tissue. To achieve this growth, dividing cells must assimilate raw materials from food. As there is a large amount of cell division in children and energy is required to synthesise new structures, the metabolic rate of children is much higher than that of adults.

The pituitary gland in the brain is in overall control of growth during childhood (Fig. 5). The anterior pituitary gland secretes a hormone called **pituitary growth hormone (PGH)**. PGH stimulates the growth of body tissues, and the elongation of the long bones leads to an increase in height. PGH binds to specific receptors on the membranes of target

cells, in these regions of the body. Genes are activated that increase the rate of assimilation of amino acids for protein synthesis, and increase the rate of mitosis. The result is growth. The pituitary also produces **thyroid stimulating hormone (TSH)**, which stimulates the **thyroid gland** to produce a hormone called **thyroxine**. Thyroxine stimulates an increase in metabolic rate and glucose metabolism, providing materials for growth.

Negative feedback is involved in controlling of the levels of these hormones; this means that an increase in the level of a hormone brings about changes that result in a decrease in its production (Fig. 6).

6a Describe the part played by the pituitary gland in growth during childhood.

b Describe how the level of thyroxine is controlled.

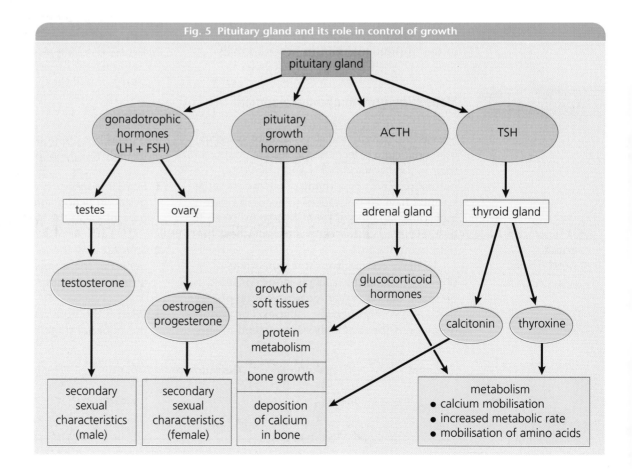

Fig. 5 Pituitary gland and its role in control of growth

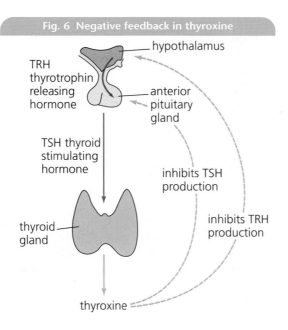

Fig. 6 Negative feedback in thyroxine

hypothalamus

TRH thyrotrophin releasing hormone

anterior pituitary gland

TSH thyroid stimulating hormone

inhibits TSH production

inhibits TRH production

thyroid gland

thyroxine

Adolescence

Childhood is a time for physical growth and learning, but then the body must undergo further changes in preparation for the reproductive phase of life. The change from child to adult is called **adolescence**. The physical changes that take place during adolescence are called **puberty**. This phase of life also includes social and emotional development when individuals become more independent of their parents. During adolescence, there is a significant increase in the growth rate in both sexes, although the timing and size of this growth spurt in boys and girls differ considerably (Fig. 1, p 51).

At puberty, the pituitary gland begins to produce the gonadotrophic hormones, luteinising hormone (LH) and follicle stimulating hormone (FSH). These hormones stimulate the testes in males and the ovaries in females. The testes produce the sex hormone, testosterone. This powerful growth promoter stimulates the development of the male secondary sexual characteristics, and initiates sperm production (Table 2). The ovaries in females produce oestrogen, which is responsible for the female secondary sexual characteristics, and is involved in the control of the menstrual cycle (Chap. 4). Oestrogen has little effect on bone and muscle growth, and the growth spurt in females is attributed to increased PGH plus hormones made by the adrenal gland. In both sexes during the growth spurt, positive feedback results in progressively larger amounts of the growth and sex hormones.

The rapid changes in growth and development that begin at puberty are probably initiated by a combination of genetic triggers and **environmental factors**, which continue to influence growth and development during adolescence. Genes determine the activity of endocrine glands, metabolism, hormonal control, and the sensitivity of the tissues to growth hormones. Many environmental factors have a significant effect on growth by acting on genes or metabolic processes. For example, a poor diet or illness during childhood could impair growth. Final adult height is achieved when the ends long bones (epiphyses) stop growing. This marks the end of adolescence.

Studies of groups of males and females show that there is considerable individual variation in growth and development for a given **chronological age**. However, on average, there is a trend over decades for

Secondary sexual characteristic	Male	Female
hair growth	on face, chest, in groin and armpits	in groin and armpits
voice	larynx enlarges greatly and voice deepens	larynx enlarges slightly
body shape	muscles and bones grow	pelvis becomes broader, fat deposited on hips and thighs, breasts develop
secondary sex organs	penis, scrotum, and prostate enlarge	uterine tubes, uterus and vagina enlarge; uterine and vaginal linings thicken
	sperm formation begins	ovulation and menstruation begin
psychological	feelings and sexual drives associated with adulthood begin to develop	feelings and sexual drives associated with adulthood begin to develop

Table 2 Secondary sexual characteristics

faster childhood and adolescent growth rate, increased adult height, and progressively earlier physical sexual development. For example, since the 1940s, there has been an increase in average adult height of 1.5 cm per decade, with individuals reaching their maximum height much earlier than previously. This could be due to better nutrition and warmer living and working conditions.

7a Explain why rapid growth occurs during puberty in males.

b Describe how the timing of puberty and the growth rate differ in males and females.

c How could better nutrition and warmer living conditions increase the adolescent growth rate and final adult height?

APPLICATION

Role of the thyroid

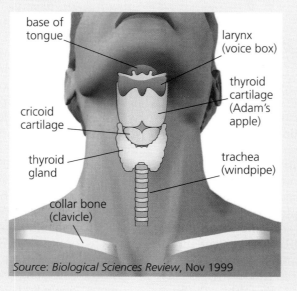

base of tongue

larynx (voice box)

thyroid cartilage (Adam's apple)

cricoid cartilage

thyroid gland

trachea (windpipe)

collar bone (clavicle)

Source: Biological Sciences Review, Nov 1999

The thyroid gland weighs about 20 g and produces thyroxine. Thyroxine binds with specific receptors on the surface of all cells of the body. It stimulates cell metabolism and mitochondrial enzymes that affect the rate of cell respiration.

The condition in which the thyroid is under-active is called **hypothyroidism**. It is often caused by a dietary lack of iodine, which is necessary to make thyroxine. A swollen neck, known as a goitre, develops as the thyroid gland makes more cells in an attempt to increase the output of thyroxine. In young children, an under-active thyroid is serious. Growth is retarded, and mental development impaired. This condition used to be known as 'cretinism'. Treatment for under-active thyroid glands is simple. Thyroid extracts were first used to treat a woman with an under-active thyroid in 1891. Since 1949, a soluble thyroxine extract has been

available in tablet form. Today, new-born babies are routinely screened to check thyroxine levels. Iodine supplements prevent goitre and most table salt is iodised.

If the thyroid is over-active, too much thyroxine is produced; this condition is called **hyperthyroidism**. Symptoms of hyperthyroidism include a fast heart rate, weight loss, and nervous or irritable behaviour. Sleeplessness and trembling hands may occur. Over-active thyroid glands can be controlled by targeting them with a controlled dose of radioactive iodine. This preparation destroys thyroid cells and thus reduces the activity of the thyroid. Doctors can then adjust the level of thyroxine in the blood with carefully measured doses of thyroxine, if necessary.

1 Thyroxine increases oxygen consumption of cells. Which cell process is affected by thyroxine?

2 Explain why an under-active thyroid is a problem in childhood.

3 Name the element essential for thyroxine production, and name two sources of this mineral.

4 a Explain how radioactive iodine is used to reduce the activity of an overactive thyroid gland.

b Why might carefully measured doses of thyroxine need to be given each day?

This woman has a goitre.

Investigating ageing

Studies have investigated growth and development in childhood and adolescence, and changes associated with senescence in adulthood. There are two types of study that can be used to assess the ageing process:

- a **cross-sectional study** looks at a groups of people of different numerical ages;

- a **longitudinal study** follows a group of people through their lives, as they age.

Cross-sectional studies involve large samples of people at several ages. Measurements of physical and physiological features can then be averaged and plotted against chronological age. Such studies are relatively quick and show general trends, but do not allow for individual lifestyles. One problem is that people in the sample may not be ageing at the same rate – they may have different **physiological ages**. Another problem is that taking a mean tends to reduce variation within the sample. This can make the rapid changes that occur in individuals at certain stages in their life, such as puberty, much less obvious. For example, individual growth rates follow a very similar pattern during adolescence, but differ considerably in their timing. A mean growth rate for a particular age range, typical in cross-sectional studies, tends to reduce the appearance of large individual changes (Fig. 7).

Longitudinal studies involve small samples and take very much longer than cross-sectional studies. However, longitudinal studies can be used very effectively to measure changes in individuals with time. For example, the effects of factors such as diet or illness can be monitored and linked to specific

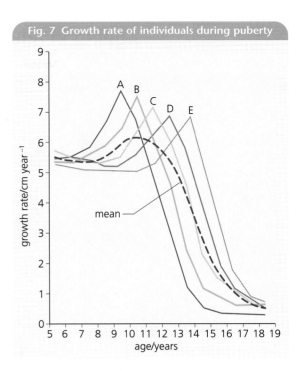

Fig. 7 Growth rate of individuals during puberty

measurements that assess the ageing process.

In both types of study, care must be taken in selecting samples. Climate, social factors, and the nature of the subject's employment could cause variation.

8a Explain how you would plan a cross-sectional study to investigate the change in height of males and females with age.

b Suggest one advantage and one disadvantage of a longitudinal study to investigate change in height with age.

KEY FACTS

- The life of a human can be divided into a pre-pubertal phase, adolescence, and adulthood.

- Humans have a long pre-pubertal stage (extended childhood) compared with other mammals.

- Growth is a product of genetic internal factors (under hormonal control) and external factors (such as diet, disease and activity).

- In children, growth is controlled by PGH, which is produced by the pituitary gland. TSH, also produced by the pituitary gland, stimulates the thyroid to produce thyroxine, which controls metabolic rate.

- Puberty marks the change from child to adult; increasing levels of hormones stimulate growth and physiological changes such as the development of secondary sexual characteristics.

- Puberty tends to happen sooner in girls than in boys.

- Longitudinal and cross-sectional studies can be used to study the effects of growth, development and ageing

5.3   Ageing in adults

Fig. 8 Key changes of ageing

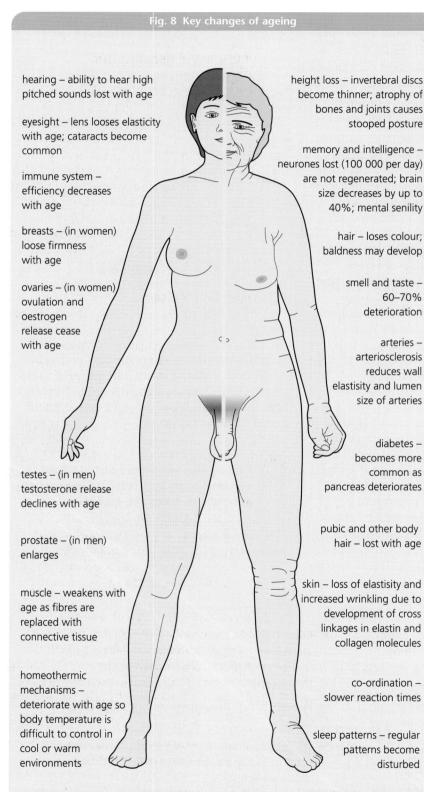

hearing – ability to hear high pitched sounds lost with age

eyesight – lens looses elasticity with age; cataracts become common

immune system – efficiency decreases with age

breasts – (in women) loose firmness with age

ovaries – (in women) ovulation and oestrogen release cease with age

testes – (in men) testosterone release declines with age

prostate – (in men) enlarges

muscle – weakens with age as fibres are replaced with connective tissue

homeothermic mechanisms – deteriorate with age so body temperature is difficult to control in cool or warm environments

height loss – invertebral discs become thinner; atrophy of bones and joints causes stooped posture

memory and intelligence – neurones lost (100 000 per day) are not regenerated; brain size decreases by up to 40%; mental senility

hair – loses colour; baldness may develop

smell and taste – 60–70% deterioration

arteries – arteriosclerosis reduces wall elasticity and lumen size of arteries

diabetes – becomes more common as pancreas deteriorates

pubic and other body hair – lost with age

skin – loss of elasticity and increased wrinkling due to development of cross linkages in elastin and collagen molecules

co-ordination – slower reaction times

sleep patterns – regular patterns become disturbed

Source: adapted from Gadd, *Individuals and Population*, Cambridge Social Biology Topics, Cambridge University Press, 1983

As we get older, our physical appearance alters due to changes in body structure that affect our external features. Internally, ageing alters the way in which organs function, and their efficiency declines from early adulthood.

Decline in physiological function

Ageing through adulthood brings a decline in many physiological functions, for example a decrease in cardiac output and a reduction in the effectiveness of gaseous exchange. Studies show that the rate of change varies between different organs (Table 3).

Senescence is the biological term for the later stages of the ageing process in late adulthood. By this time, decline in function has significant effects on body functions and abilities. Both appearance and body function change substantially (Fig. 8).

9a   Describe how ageing may cause changes in the external appearance of an individual.

b   Which physical changes are specific to males and to females?

c   Older people find it more difficult to sustain physical activity. Use Table 3 and Fig. 8, and your knowledge of physiology from A/S, to explain why.

Table 3 Physiological function of men aged 75 compared with men aged 30	
Function	Mean % of function remaining
nerve conduction velocity	89
maximum oxygen uptake (during exercise)	41
basal metabolic rate	86
kidney filtration rate	68
cardiac output	70
vital capacity (maximum lung volume)	50

Interaction of genetic and environmental factors

Old age or senescence is the result of the interaction between genetic and environmental factors.

Some genetic factors

Sufferers from Werner's syndrome show premature ageing. They have grey hair and wizened skin by their early 20s, and often die before age 50. The cause of this condition is the failure of an enzyme called helicase to remove damage or breakages in unwound DNA following transcription. These irreversible changes in DNA in cells of the body are called **somatic mutations**. Mistakes in DNA normally accumulate slowly during a lifetime and lead to ageing, but in Werner's syndrome the process is accelerated.

When cells divide repeatedly, the ends of chromosomes progressively shorten. The loss of the ends, called telomeres, is associated with changes in cell activity and a reduction in cell division in the affected cells.

Ageing in skin cells produces collagenase, an enzyme that alters collagen structure and is associated with wrinkles.

Some environmental factors

There are a number of environmental factors that can affect body tissues. Tissues in contact with the external environment are most at risk. Substances in the food we eat, ultraviolet light, air pollutants, exposure to chemicals in the workplace or the home, and pathogens all affect body tissue (Chap. 8). By limiting the frequency and amount of exposure to these factors, the ageing process might be slowed down.

Interaction of genetic and environmental factors

There are three ways to explain how genetic and environmental factors might interact in the ageing process:

- accumulation of genetic error;
- degeneration of tissues;
- malfunction of the immune system.

Accumulation of genetic error

Somatic mutations can be caused by environmental factors and occur in the non-reproductive cells of an organism. These mutations are passed on during mitosis, when new cells are produced for growth and repair of tissues. Not only is each mistake copied, but additional mutations may occur and also be passed on during subsequent cell divisions. Errors in the DNA can lead to incorrect cell function. The accumulation of numerous genetic errors leads to progressively more and more malfunctioning tissue, and results in some of the features of ageing.

Degeneration of tissue

Degeneration of tissue is largely due to 'wear and tear' and incorrect repair of damaged tissue. For example, changes in tissue elasticity lead to changes in our outward appearance. Similarly, changes in the structure of internal organs lead to reduced efficiency. Exposure to substances in the environment can accelerate this degeneration. For example small particles from diesel emissions called PM10s (less than 10 micrometres across), damage the delicate lung surface; ultraviolet light affects exposed skin by thickening the epidermis and making it tougher; components of the diet affect the alimentary canal or may be absorbed and transported to tissues via the bloodstream. Injury, or bad living or working conditions may accelerate wear and tear.

Malfunction of the immune system

The immune system recognises 'self' (the body's own proteins) as different from 'non-self' (other proteins such as those of invading bacteria or viruses). So, the immune system can protect us from harmful microorganisms by recognising and destroying them (Chap. 7) It also protects us from our own tissues when things have gone wrong, for instance, when a cell has developed cancerous features and begins to divide in an uncontrolled way. As we age, our immune system becomes less efficient. The ability to recognise and destroy invading pathogens or cancerous cells declines.

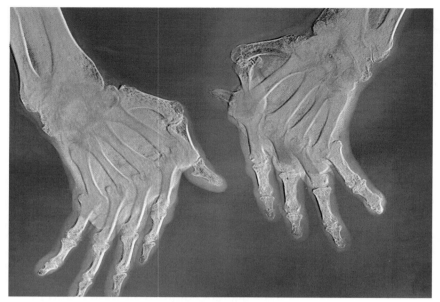

Autoimmune response in the joints of the hands and feet causes rheumatoid arthritis. In this false-colour X-ray of the hands of a person with extreme rheumatoid arthritis, the joints of all the fingers of both hands are severely damaged.

Endothelial cells produce a substance called interleukin-1 in response to irritants or pathogens. By over-reacting to the presence of foreign substances, endothelial cells lining blood vessels, the gut, and airways in the lung may cause damaging inflammation in these parts of the body. As a result, function is impaired.

The immune system may also destroy the body's own healthy tissue by mistake. This is called **autoimmunity**. It is responsible for the diseases rheumatoid arthritis and multiple sclerosis.

10 Briefly summarise the part played by genes and by the environment in:

a accumulation of genetic error;

b degeneration of tissues;

c malfunction of the immune system.

KEY FACTS

■ Ageing includes all the growth, development and physiological changes that occur in a lifetime.

■ Deterioration in external appearance and decline in function of body organs with age is called senescence.

■ Decline in function associated with ageing is caused by interactions between genetic and environmental factors. It involves the accumulation of genetic error, degeneration of tissue, and malfunction of the immune system.

EXAMINATION QUESTIONS

1 The table shows the timing of some of the aspects of puberty in a large sample of girls and boys.

Event	Average age at which event begins/years	Range of ages at which event begins/years	Average age at which event ends/years	Range of ages at which event ends/years
height spurt in girls	10.5	8.5–14.0	14.0	12.5–15.5
development of breasts	10.8	8.0–13.0	14.8	12.0–18.0
first menstrual period	13.0	10.5–15.5	not applicable	not applicable
height spurt in boys	12.5	10.5–16.5	16.0	14.0–17.5
growth of penis	12.5	10.5–14.5	14.5	12.5–16.5
growth of testes	11.5	9.5–13.5	15.5	14.0–17.0

a Suggest **two** possible explanations for the variation in the age at which puberty begins in girls. (2)

b Use the information in the table to give:
  i) the earliest age at which a girl in the sample could have completed the aspects of puberty shown; (1)
  ii) the range of ages when **all** boys in the sample were in the process of puberty. (1)

AQA/NEAB BY09 June 1996 Q5

2 The graph shows growth curves of different parts of the human body expressed as a percentage of post-natal growth.

a Between which ages does the most rapid growth of the whole body occur? (1)

b i) Describe the pattern of growth of the brain and head.

ii) Suggest the importance of this pattern of growth to a human. (3)

c i) Describe the pattern of growth of the reproductive organs.

ii) Suggest the importance of this pattern of growth to a human. (3)

AQA/NEAB BY09 June 1997 Q2

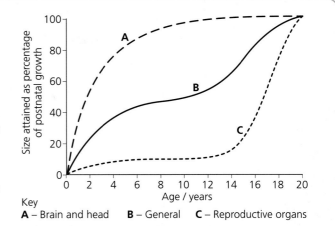

Key
A – Brain and head   **B** – General   **C** – Reproductive organs

3

a Many older people find it difficult to run to catch a bus. Describe **three** of the physiological changes which occur with ageing that may contribute towards this. (6)

b Explain how changes in each of the following may contribute to the ageing process:

i) genes;
ii) tissues;
iii) the immune system. (6)

AQA/NEAB BY09 Q9

4

a Monozygotic twins are sometimes called identical twins. However, there are slight differences in the appearance of monozygotic twins at birth. Explain why. (2)

b A twin study carried out in the United States investigated the effects of genes and environment on the length of life. The investigators recorded the differences in age at death for each pair of twins and calculated mean values. The results are shown in the table.

i) The researchers only collected data from twin pairs in which the first twin to die was over 60 years old. Explain why. (1)

ii) Use the data to evaluate the effects of genes and environment on the length of life. (3)

	Monozygotic twins		Dizygotic twins	
	female	male	female	male
Mean difference in age at death/months	47.6	29.4	89.1	61.3

Hong Kong is a very prosperous and rapidly changing community.

We live in a rapidly changing world. With new technology, humans have made enormous progress in their ability to communicate, travel, conquer disease, and provide food for increasing populations.

With a greater life expectancy and improved medical provision, birth rates exceed death rates in most parts of the world. Many of the factors that limit the growth of other animal populations have been reduced, so that the world human population now increases by around 80 million per year. With more mouths to feed, agriculture must become more productive, but expanding towns put additional pressures on land for housing, industry, and roads. At the start of the third millennium, most people live in or around towns. Lack of space means many cities have grown upwards as well as outwards.

In general, population growth in the more economically developed nations (MEDNs) is slower than in the less economically developed nations (LEDNs) and is likely to remain so, as the graph below left shows.

Predictions of population growth, such as the pie chart below right, show most growth occurring in LEDNs.

There are now serious concerns about the growth of human populations. Continued expansion will not only affect the quality of human life, but more importantly, may destroy the planet itself.

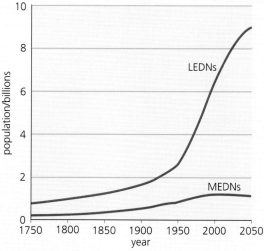

Source: United Nations, *World Population Prospects: The 1998 Revision (1998)*; estimates by the Population Reference Bureau

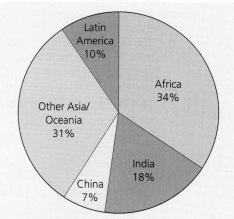

Source: United Nations, *World Population Prospects: The 1998 Revision (1998)*; estimates by the Population Reference Bureau

6.1 Population size

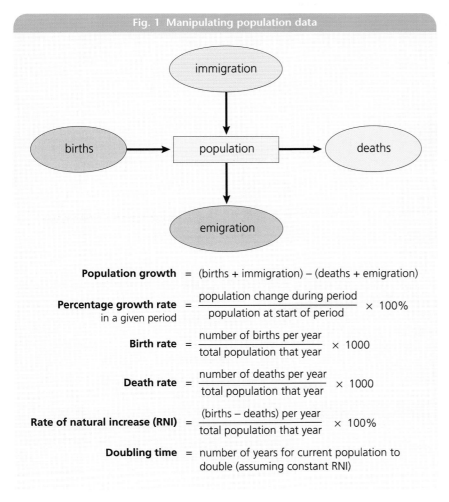

Fig. 1 Manipulating population data

Population growth = (births + immigration) – (deaths + emigration)

$$\text{Percentage growth rate in a given period} = \frac{\text{population change during period}}{\text{population at start of period}} \times 100\%$$

$$\text{Birth rate} = \frac{\text{number of births per year}}{\text{total population that year}} \times 1000$$

$$\text{Death rate} = \frac{\text{number of deaths per year}}{\text{total population that year}} \times 1000$$

$$\text{Rate of natural increase (RNI)} = \frac{(\text{births} - \text{deaths}) \text{ per year}}{\text{total population that year}} \times 100\%$$

Doubling time = number of years for current population to double (assuming constant RNI)

A **population** is defined as the number of individuals of a species in a specified area. The total world human population exceeds 6 billion. Most countries keep accurate records, so that their total population can be determined. These records include:

- births;
- deaths;
- immigration (people entering the country);
- emigration (people leaving the country).

There are several useful ways of examining population data (Fig. 1).

1a Use Table 1 to calculate the RNI for the UK and Hong Kong.

b Which would have the shortest doubling time?

c Explain why RNI figures may be an inaccurate measure of the growth rate for a population.

Table 1 Population data for the UK and Hong Kong, 1999			
Country	Total population	Births (per 1000) per year	Deaths (per 1000) per year
UK	59.4 million	12	10
Hong Kong	6.9 million	9	5

KEY FACTS

- Population growth is defined as (births plus immigration) minus (deaths plus emigration).

- Growth rate of a population can be shown as percentage increase per year, known as the rate of natural increase (RNI).

6.2 Population structure

The standard of living in individual countries varies enormously and affects population structure. In MEDNs (for example, the United States, Western Europe and Australia), the majority of the population has high living standards, plenty of food, and good medical services. This may contrast sharply with the conditions in LEDNs (for example, sub-Saharan Africa). Beyond the relatively few more affluent centres, the majority of the population in LEDNs are the rural poor.

Population pyramids give a visual representation of the age structure of a population. The percentages of males and females in each age group of a population are represented by horizontal bars, males on the left females on the right. The structures of populations reflect the differences in health

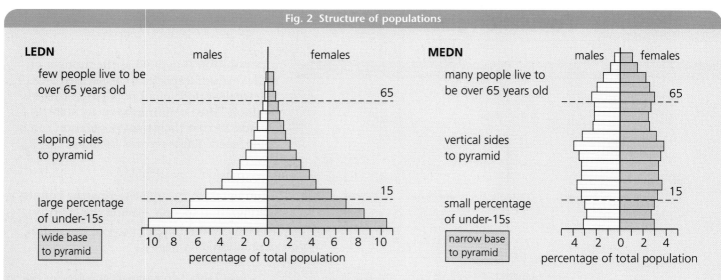

Fig. 2 Structure of populations

LEDN

few people live to be over 65 years old

sloping sides to pyramid

large percentage of under-15s

wide base to pyramid

males — females

65

15

10 8 6 4 2 0 2 4 6 8 10
percentage of total population

MEDN

many people live to be over 65 years old

vertical sides to pyramid

small percentage of under-15s

narrow base to pyramid

males — females

65

15

4 2 0 2 4
percentage of total population

Source: adapted from Hornby and Jones, *Introduction to Population Geography*, Cambridge University Press, 1993

caused by food availability, medical provision, and housing and working conditions. These factors affect the number of births and the pattern of deaths in each age group. So, the population pyramids of an MEDN and an LEDN show characteristic differences in three ways:

- the angle of the sides;
- the height;
- the width of the base (Fig. 2).

LEDNs usually have both a high birth rate and high **infant mortality rate**. Typically, over 35% of the population are children under 15 years of age, so the base of the population pyramid is broad. Death rates for adults are high and few survive to old age, so the pyramid is short in height. Children under 15 and adults over 65 are usually dependent on the working population in between. It is useful to examine the proportion of people under 15 and over 65 in a population. A large under-15 group is typical in an expanding population, and a large proportion of over-65s is more typical in a developed country.

A **survival curve** is a graph plotted from records of the actual lifespans of individuals in a sample group of 10 000 (Fig. 3). The **average life expectancy** is the age at which 50% of the people in the sample are still alive. So, average life expectancy can be calculated from a survival curve.

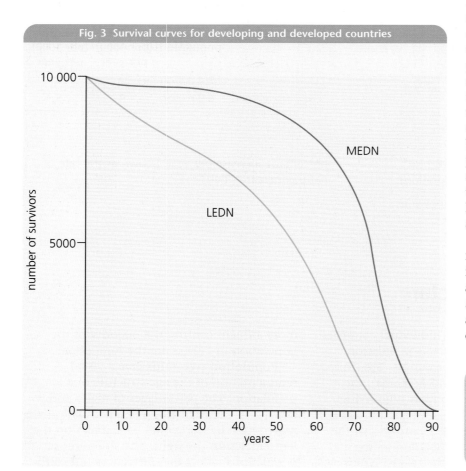

Fig. 3 Survival curves for developing and developed countries

number of survivors

10 000

5000

0

MEDN

LEDN

0 10 20 30 40 50 60 70 80 90
years

2a Use Fig. 3 to calculate average life expectancy in an LEDN and an MEDN.

b Suggest the three most important factors that would improve the average life expectancy of a group of people.

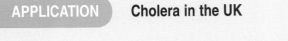

APPLICATION   **Cholera in the UK**

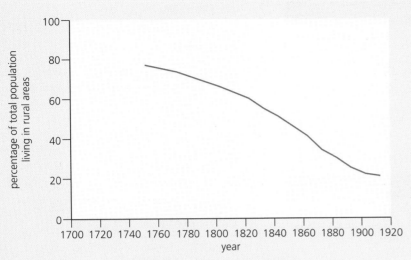

Source: based on data in Grigg, *Population Growth and Agrarian Change*, Cambridge University Press, 1980

Rural population in England and Wales 1700–1920.

Sewage systems were rebuilt in all large towns in the late 1800s and many, like these sewers in Brighton, are still in use today.

As the graph on the left shows, the percentage of people living in rural areas in England and Wales began to decline in the mid-1700s and the population became increasingly town-based. This is known as urbanisation. The rapid development of urban communities was linked to industrialisation and world trade. Towns offered better opportunities for employment, and working people could earn more money. Imported food reduced the dependence of town-dwellers on the farming communities and farm wages dropped. This accelerated the rural decline.

Outbreaks of cholera were frequent in the UK in the mid-1800s. In 1817, a cholera pandemic started in Bengal; it reached Moscow by 1830, and England by 1831. It was probably brought into Sunderland by infected passengers on a ship from Europe. Once here, the disease spread rapidly through the UK claiming 50 000 to 70 000 lives.

John Snow, then a medical apprentice, noticed that the disease did not just strike the poor and undernourished, although many more of them died than the wealthy. Snow carried out detailed studies and concluded that:

- the sick passed the disease to their carers;
- cholera spread quickly in overcrowded houses where up to 15 families shared facilities;
- the source of drinking water was the main link between sufferers;
- fresh water from street pumps or wells was contaminated by waste from cracked sewer pipes and this was the method of transmission.

Cholera was eradicated from the UK largely because of improvements in water treatment and distribution. Better housing with sanitary provision, a reduction in overcrowding, and simple personal hygiene were also important.

1 Explain why cholera epidemics have their greatest effects on urban populations.

2 Describe two contributions to the eradication of cholera from the UK.

APPLICATION   **Life expectancy – past and present**

Life expectancy changes with time and conditions					
Average life	London	Liverpool	Manchester	Surrey	Glasgow
1841					
males	35	25		44	
females	38	27		46	
1881					
males			29	51	35
females			33	54	44

Average life expectancy around the world in 1999				
Continent	Region	Average life expectancy for the region/years	Range of average life expectancy/ years*	GNP *per capita* (US$)
Africa	Northern Africa	64	46–68	1160
	Western Africa	53	44–65	340
	Eastern Africa	44	37–74	260
	Middle Africa	49	46–64	300
	Southern Africa	56	40–58	3030
North America		77	76–79	28130
Latin America	Central America	71	65–76	3950
	Caribbean	69	54–78	3300 (est.)
	South America	69	60–74	4430
Asia	Western Asia	68	59–77	3370
	South Central Asia	61	46–73	470
	South-east Asia	65	51–77	1610
	East Asia	72	63–81	4580
Europe	Northern Europe	77	70–79	21500
	Western Europe	77	72–79	27900
	Eastern Europe	69	67–74	2510
	Southern Europe	77	72–78	15480
Australasia		74	56–78	16460

* Average life expectancy varies from country to country within region

1 Study this table.

a Explain why the life expectancy differed between Surrey and the cities.

b Suggest two reasons why it is difficult to draw a firm conclusion from this data.

2 Study this table.

a Suggest which countries you would classify as MEDNs and LEDNs.

b Does there seem to be a link between income (*per capita*) and average life expectancy?

c Calculate a Spearman's rank value for income (*per capita*) and average life expectancy. Use the formula:

$$r_s = 1 - \frac{6\Sigma d^2}{n(n^2 - 1)}$$

where Σ = sum of
    d = difference between the GNP and life expectancy rank values
    n = number of samples.

d Comment on the statistical significance of the correlation. Use the chart or table on p. 71.

KEY FACTS

- The changes in a population's size and structure can be represented by the demographic transition model.

- Food supply had a major effect on the growth of the population in Ireland in the 18th century.

- The control of water-borne disease, reduced mortality rates in early industrial societies.

- Vaccination has reduced the number of deaths caused by some infectious diseases.

EXAMINATION QUESTIONS

1 The graph shows the changes in birth and death rate in England and Wales during the eighteenth and nineteenth centuries.

a i) During which year did the population grow least? (1)

ii) Calculate the population growth rate in 1861. Show your working. (2)

b Two reasons for the change in death rate between 1741 and 1901 were the introduction of vaccination and the effective disposal of sewage. Explain how each of these two factors was effective in reducing the death rate.

i) vaccination (2)

ii) sewage disposal (2)

AQA/NEAB BY09 June 1996 Q6

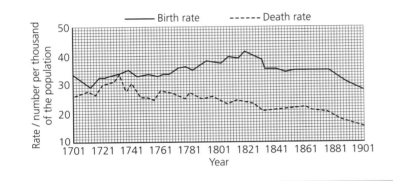

2 The graph shows changes in birth rate and in death rate at various stages in the human demographic transition.

a Using specific examples to illustrate your answer, describe the importance of disease and food supply at stage A. (6)

b What are the main ways in which the structure of the population differs in stages A and B? (6)

AQA/NEAB BY09 June 1995 Q9

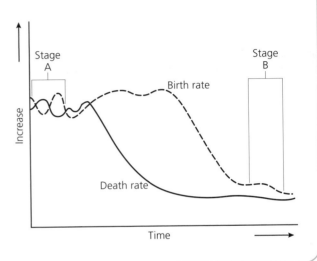

REFERENCE MATERIAL

Spearman's rank correlation coefficient

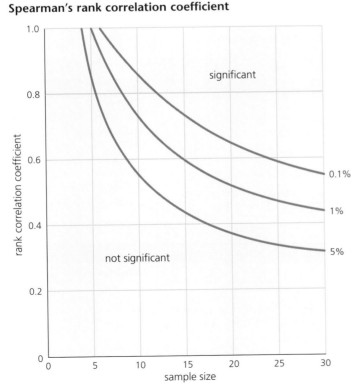

Critical values table for Spearman's rank coefficient at $p = 0.05$ level	
No. of pairs of measurements	**Critical value**
5	1.00
6	0.89
7	0.79
8	0.74
9	0.68
10	0.65
12	0.59
14	0.54
16	0.51
18	0.48
20	0.45
22	0.43
24	0.41
26	0.39
28	0.38
30	0.36

7 Infectious disease

These young people can expect a longer, healthier life than their parents.

Widespread outbreaks of disease are called **epidemics**, and epidemics that spread internationally are called **pandemics**. Pandemics of the 'black death' in pre-industrial Europe reduced the population by 30–50% in the 14th century. Nowadays, in MEDNs, clean drinking water, sewage treatment, and high standards of food hygiene help prevent the spread of disease. Immunisation provides protection from many infectious diseases, and health services provide expertise and drugs in the event of illness. The average life expectancy in these nations has risen.

In overcrowded conditions, infectious diseases spread rapidly and contribute to a high death rate. But the spread of infectious disease is neither a problem of the past, nor limited to LEDNs today. Influenza pandemics regularly kill people all over the world. Other communicable diseases currently causing concern include HIV infection, and a virulent form of the bacterium that causes TB.

7.1 What causes infectious diseases?

Fig. 1 Biology of viruses, bacteria and fungi		
Organism	**Characteristic features**	**Pathogen and disease**
Viruses lipid envelope, neuraminidase, haemagglutinin, helical nucleocapsid consisting of RNA, protein matrix — 150 nm	• small (ultra-microscopic) 20–3000 nm but typically they are 100–300 nm • acellular • obligate intracellular parasites • ability to reproduce only within the host cell • reproduction exploits metabolism and materials of host cell	influenza virus – influenza mumps virus – mumps human immunodeficiency virus (HIV) – acquired immune deficiency syndrome (AIDS) rubella virus – german measles measles virus – measles
Bacteria mesosome, stored food reserve, capsule, cell wall, cell membrane, ribosomes, plasmid, nuclear material – a loop of DNA — 1 μm	• small (0.5–10 μm) • prokaryotic (lack nuclei, nucleic acid not membrane-bound) • heterotrophic nutrition • asexual reproduction by binary fission • sexual reproduction by transfer of genetic material from one cell to another	*Salmonella enteritidis* – Salmonella food poisoning *Mycobacterium tuberculosis* – tuberculosis (TB) *Corynebacterium diphtheriae* – diphtheria *Clostridium tetani* – tetanus *Bordetella pertussis* – pertussis (whooping cough)
Fungi endoplasmic reticulum, cell wall, mitochondrion, nucleolus, vacuole, nucleus, double nuclear membrane, cell membrane, hyphae — 1 μm	• thread-like growth (hyphae) in filamentous forms of fungi • cell walls made of chitin • eukaryotic with nucleus and membrane-bound organelles • heterotrophic nutrition • reproduction by spores	*Candida* – thrush *Epidermophyton* – athlete's foot *Epidermophyton* – ringworm

Communicable diseases are caused by microorganisms, commonly bacteria, viruses, and fungi (Fig. 1).

Infection occurs when microorganisms get past the external defence mechanisms and enter the body tissues. Infection may occur through cuts or abrasions in the skin, and via natural openings of the breathing, digestive and urino-genital systems. Not all infections cause **disease**. The microorganisms that cause disease are known as **pathogens**.

Pathogens colonise and reproduce in tissues and body fluids causing physical damage to cell structure, disrupting cell functions, releasing toxins, and stimulating the response of the body's immune system (Fig. 2). These actions produce the **signs and symptoms** we call disease. The signs of a disease are the visible features of the disease. For example, somebody suffering from

measles will develop a rash and a fever with a high temperature. The sufferer of a disease also experiences the symptoms of the disease, which are not visible to another person. For example, somebody suffering from measles may feel nauseous and very weak, or have various aches and pains.

For an infection to take hold and cause disease, an organism must do three things:

- attach itself to host tissues;
- penetrate the host cells;
- colonise and reproduce within the host tissue.

Pathogens can recognise and attach to host cells. Receptor binding protein molecules, called **ligands**, are found in the microbial wall or viral coat. These bind with specific protein receptor molecules on the host cell membrane. Many organisms may bind to cells but do not enter. Pathogens enter host cells by endocytosis or by producing enzymes that breach the host cell membrane. Once inside, colonisation depends on the ability of the pathogen to reproduce in the host tissues. This may take time, and the period of time between infection and the appearance of the sings and symptoms of a disease is called the **incubation period**.

It is possible for an infected, but otherwise apparently healthy individual, to infect another person. The person passing on the infection, but showing no signs or symptoms of disease, is called a **carrier**. A person may be carrier during the incubation period, but in some cases the disease may never develop. For example, the bacterium *Mycobacterium tuberculosis* that causes tuberculosis (TB) is thought to be carried by about 1.9 billion people at any one time. Only a small percentage (about 20 million people) actually develop TB, so there is an immense disease reservoir carried by people who are unaware that they are infected.

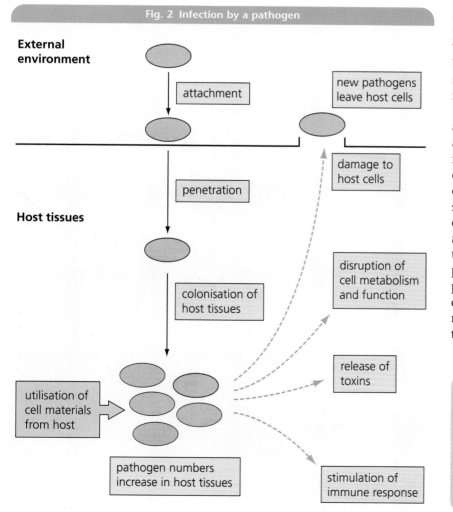

Fig. 2 Infection by a pathogen

External environment

attachment

new pathogens leave host cells

Host tissues

penetration

damage to host cells

colonisation of host tissues

disruption of cell metabolism and function

utilisation of cell materials from host

release of toxins

pathogen numbers increase in host tissues

stimulation of immune response

1a What is a pathogen?

b Distinguish between the terms infection and disease.

c Explain the term carrier.

d What percentage of TB carriers develop the disease?

7.2 Transmission of pathogens

Pathogens can infect an individual in a number of ways (Fig. 3). Transmission can be by:

- air;
- contaminated water;
- contaminated food;
- direct contact (including sexual intercourse);
- animals (vectors).

Air-borne infection

Many pathogens are air-borne (Table 1). When an infected person coughs, sneezes, talks, or breathes out, microorganisms are passed into the atmosphere in droplets of saliva, mucus and water. Large droplets tend to travel only a metre or two before reaching the ground, but smaller droplets can remain suspended for long periods. Droplets that land on the ground dry out fairly quickly and expose the microorganisms to air currents that can waft them around a room. In unventilated crowded places, such as buildings or public transport, air may carry a large number of pathogens. Dust particles and flakes of dead skin also transport microorganisms, a significant factor in infection in hospital wards.

Ventilation systems can recycle pathogen-laden air round buildings, and the more confined space in passenger aircraft. Fresh air entering a building can contain pathogens and sometimes the pipes in ventilation systems become reservoirs of pathogenic microorganisms. This can lead to 'sick building syndrome'. In 1994, the Inland Revenue closed one of its Merseyside offices because over half of its of two thousand employees had suffered repeated bouts of flu and other air-borne diseases over a five-year period. Rather than replace the ventilation system, the office was moved.

2 Suggest two ways that the risk of air-borne infection is reduced in operating theatres.

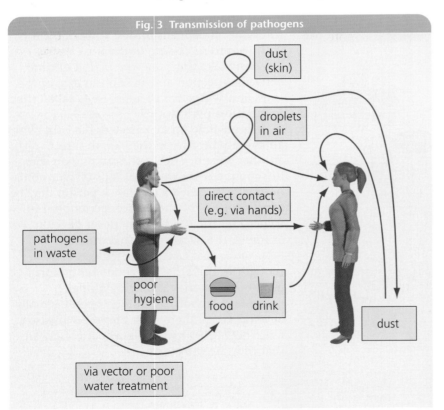

Fig. 3 Transmission of pathogens

dust (skin)

droplets in air

direct contact (e.g. via hands)

pathogens in waste

poor hygiene

food   drink

dust

via vector or poor water treatment

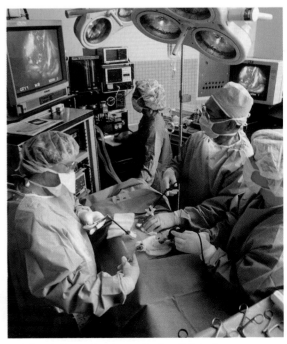

Air in hospital operating theatres is often slightly above atmospheric pressure to prevent air from other areas entering. Staff wear masks to filter out pathogens from breath. The patient is covered with sterile clothes to prevent air-borne infection.

Table 1 Common air-borne diseases	
Viral diseases	**Bacterial diseases**
influenza (flu)	tuberculosis
common cold	sore throat caused by *Streptococcus*
rubella	

Influenza

Influenza virus cannot survive for more than a few hours outside a living cell. The virus attaches to specific receptors on the surface of epithelial cells in the nose, throat and bronchi. The host cell membrane folds around the virus to enclose it within the cytoplasm, but the virus then destroys the membrane so that it is free to enter the cytoplasm. Once inside the cell, the nucleic acid is released from the virus, and it begins to reproduce (Fig. 4). New viruses bud off from the infected cell. Cell disintegration or **lysis** follows the death of the epithelial cell when mucus production increases, and sneezing disperses the virus to infect other people. The epithelial cell damage can lead to serious secondary infections such as bronchitis or pneumonia.

After an incubation period of two or three days, a high temperature, aching limbs, and fatigue are caused by the body's defence mechanisms. Once the immune system produces antibodies, the number of virus particles begins to decline and the symptoms subside.

Water-borne infection

Diarrhoea affects over 1 billion people a year and kills over 5 million, mainly children. Almost all of the deaths are in LEDNs. Many pathogens can cause diarrhoea, including the bacteria *Vibrio cholerae* (Chap. 6) and *Escherichia coli* in water contaminated with human faeces.

In December 1992, a cholera epidemic began in Bangladesh. It was caused by the 'Bengal' strain of *V. cholerae*. This epidemic has spread to at least 11 more countries in southern Asia. The current cholera 'dead' vaccine (see pp. 80 and 81) offers only limited immunity to the disease for about three months.

Water can be tested for the presence of bacteria. The presence of large numbers of bacteria in the water, including *E. coli,* indicates that there is sewage contamination. Water from reservoirs is treated to ensure a safe supply of drinking water (Fig. 5 overleaf).

3 Reservoirs are increasingly used for water sports. Study Fig. 5 and explain how the water treatment process reduces the risk of disease-causing organisms entering the water supply.

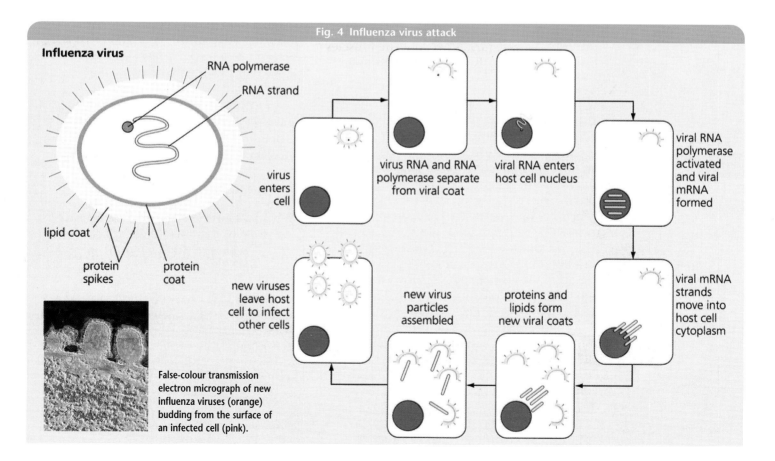

Fig. 4 Influenza virus attack

Influenza virus

RNA polymerase

RNA strand

lipid coat

protein spikes

protein coat

virus enters cell

virus RNA and RNA polymerase separate from viral coat

viral RNA enters host cell nucleus

viral RNA polymerase activated and viral mRNA formed

viral mRNA strands move into host cell cytoplasm

proteins and lipids form new viral coats

new virus particles assembled

new viruses leave host cell to infect other cells

False-colour transmission electron micrograph of new influenza viruses (orange) budding from the surface of an infected cell (pink).

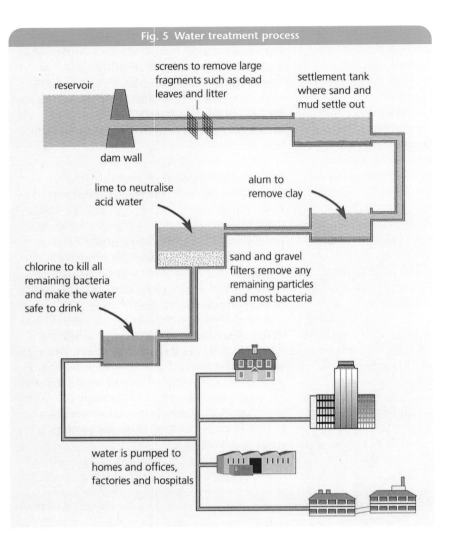

Fig. 5 Water treatment process

reservoir

screens to remove large fragments such as dead leaves and litter

settlement tank where sand and mud settle out

dam wall

lime to neutralise acid water

alum to remove clay

chlorine to kill all remaining bacteria and make the water safe to drink

sand and gravel filters remove any remaining particles and most bacteria

water is pumped to homes and offices, factories and hospitals

must be sufficient time for pathogens to reproduce. Bacteria cause most food-borne diseases, and infections usually arise when one or more of the guidelines in Fig. 6 are not followed. They all aim either to prevent contamination or to restrict microbial growth. Food retailers also have strict hygiene regulations for workers' clothing, cooked and uncooked foods must be kept separate, surfaces kept clean, and storage temperatures are strictly controlled. Bacterial diseases spread by contaminated food include *Salmonella* food poisoning, botulism, enteritis, and typhoid.

The pathogens may produce toxins that are harmful if ingested, or the organism must be present in sufficient numbers to cause infection when the food is eaten. *Clostridium botulinum* produces botulinal toxin in anaerobic conditions. This potent nerve toxin causes paralysis when eaten. *Staphylococcus aureus* can grow on salted meat and preserved milk products; it produces a toxin that causes severe vomiting and diarrhoea. The toxin will cause disease even if the bacteria are not present.

Salmonella

Salmonella enteritidis is a common cause of bacterial food poisoning (Fig. 7). The organisms are transmitted in food that is contaminated with the faeces of a carrier. The carrier may be human or animal. The symptoms of *Salmonella* food poisoning include stomach pain, nausea, vomiting and diarrhoea, chills, headaches, fatigue, and fever.

Diarrhoea

E. coli is present in faeces because it is part of the normal flora of the large intestine. However, pathogenic strains enter the epithelial cells of the gut. The pathogenic strains cause the secretion of large amounts of fluid into the digestive tract resulting in diarrhoea. The body loses salts at the same time as it loses water. People suffering from severe diarrhoea are often given oral rehydration therapy (ORT). They drink a solution of salts and sugars to replace some of the substances the body is losing. Many lives, especially those of children in LEDNs, are saved by this technique. However, the disease will recur if poor sanitation is not improved and sewage-contaminated water supplies are not cleaned.

Food-borne infection

For pathogens to cause food poisoning, conditions must be suitable for microbial growth (e.g. warm temperature), and there

Fig. 6 Some do's and don'ts of food preparation

- Don't buy food outside its sell-by date
  (*microbial growth might have begun*).
- Don't refreeze frozen food that has thawed out
  (*microbial growth will have occurred as food thawed if microbes are present*).
- Do put perishable food in the fridge
  (*cold restricts microbial growth*).
- Do wash vegetables thoroughly
  (*to remove soil microbes*).
- Do keep your food preparation area clean
  (*to reduce the risk of contaminating food*).
- Do wash your hands before handling food
  (*to prevent faecal and other contamination*).
- Do keep hot food hot and cold food cold until eaten
  (*to prevent microbial growth occurring as food reaches a moderate temperature*).
- Do be certain that you reheat thoroughly if you reheat food
  (*to kill microbes that might have started to grow*).

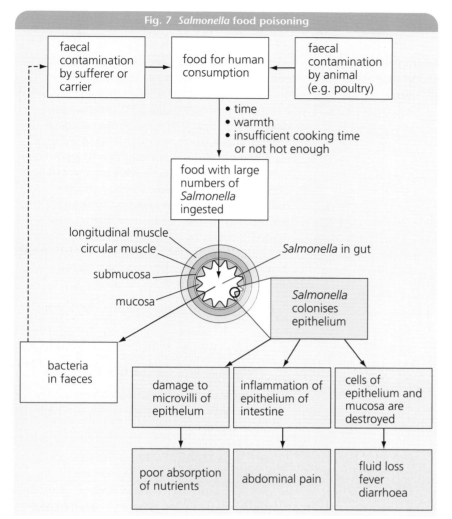

Fig. 7 *Salmonella* food poisoning

faecal contamination by sufferer or carrier → food for human consumption ← faecal contamination by animal (e.g. poultry)

food for human consumption →
- time
- warmth
- insufficient cooking time or not hot enough

food with large numbers of *Salmonella* ingested

longitudinal muscle
circular muscle
submucosa
mucosa

Salmonella in gut

Salmonella colonises epithelium

bacteria in faeces

damage to microvilli of epithelum | inflammation of epithelium of intestine | cells of epithelium and mucosa are destroyed

poor absorption of nutrients | abdominal pain | fluid loss fever diarrhoea

4 Suggest why the following help to prevent food-borne infections:

a cooking food and maintaining at high temperatures after cooking;

b washing hands frequently when handling or preparing foods;

c storing foods for short periods of time at cool temperatures;

d keeping cooked and uncooked foods separate.

Direct contact infection

When you have a cold and blow your nose, some virus particles can end up on your hands instead of in the tissue. If you shake hands with someone who then brings their hand into contact with their nose, the virus could infect the second person. Washing your hands reduces the risk of this person-to-person infection.

Sexual intercourse can also spread infection. Bacterial diseases such as syphilis and gonorrhoea, and viral diseases like genital herpes can be passed on by sexual contact.

Contaminated hypodermics and blood products increase the risk of infection by enabling pathogens to get past the body's natural barriers.

5 Explain three ways that could reduce the spread of direct contact diseases.

Athlete's foot

Athlete's foot is caused by the fungus *Epidermophyton*. It attacks damp areas of the body, like the skin between the toes. The fungus grows as thread-like hyphae through the surface of the skin, and feeds on dead epidermal cells (Fig. 8). Digested epidermal cells reveal the more delicate tissue below. The infected area becomes red, wet, itchy and sore. Spores from the fungus are shed onto floors, into clothing or onto towels. Damp changing rooms and the swimming pools are the usual places where infection occurs.

Fig. 8 *Epidermophyton* infection

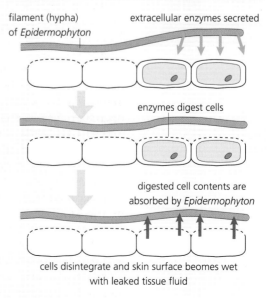

filament (hypha) of *Epidermophyton*   extracellular enzymes secreted

enzymes digest cells

digested cell contents are absorbed by *Epidermophyton*

cells disintegrate and skin surface beomes wet with leaked tissue fluid

Flu

Flu is usually transmitted by droplet infection. Damp air and lack of ventilation may allow the air to be laden with droplets containing virus particles. The graph below shows the weekly consultation rate for flu in England.

Most people who get flu recover completely in 1–2 weeks, but some people develop potentially life-threatening medical complications, such as pneumonia. Although such complications can occur at any age, elderly people and people with chronic health problems are particularly at risk. Old age reduces the immune response and older people may already be weaker due to age-related diseases or other infections.

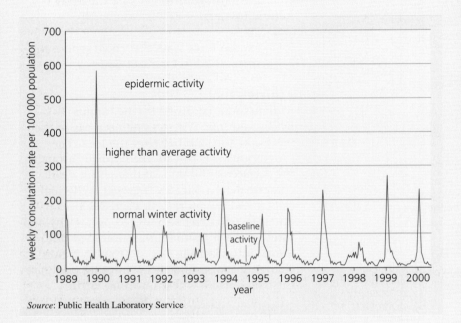

Source: Public Health Laboratory Service

1 a Suggest the time of year at which is flu most common.

b Suggest why transmission is most likely at this time.

2 a Make a bullet-point list of advice on how to reduce the risk of catching flu.

b Suggest how a health authority could ensure this advice reaches the two high-risk groups.

- Pathogens are microorganisms that cause disease.

- Infectious diseases can be passed from one person to another.

- Transmission may be by air, droplet, contaminated water and food, and direct contact.

- Understanding the mode of transmission enables control of the spread of disease.

7.3   Preventing infection

The human body has several ways to keep itself free from pathogens. The first line of defence is the presence of physical and chemical barriers that stop pathogens entering body tissues: skin, mucus hydrochloric acid, and lysozyme.

The skin is a tough dry barrier covering the body surface. The dead epithelial cells are impregnated with keratin prevent infection by microorganisms. All moist surfaces such as ducts, the lungs, gut, and reproductive tracts have a physical barrier of slimy mucus secreted by goblet cells in the epithelium. Hydrochloric acid is secreted by glands in the stomach; the stomach contents are usually about pH 2. Lysozyme is a protease. This enzyme, found in tears and saliva, prevents microbial growth by rupturing the cell walls of bacteria.

6 Describe the physical and chemical barriers to each of the following:

a droplet infection;

b water-borne and food-borne infection;

c contact infection.

A second line of defence is needed to attack pathogens that have entered the body tissues. This is the role of the immune system. To be

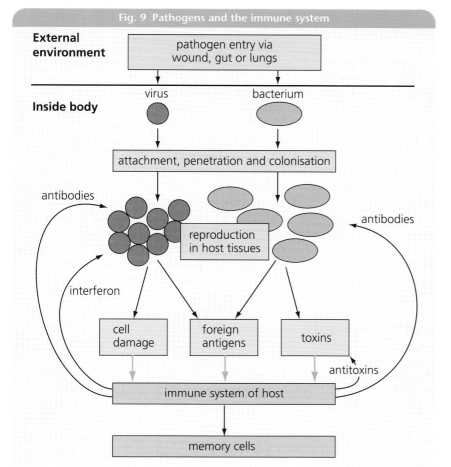

Fig. 9 Pathogens and the immune system

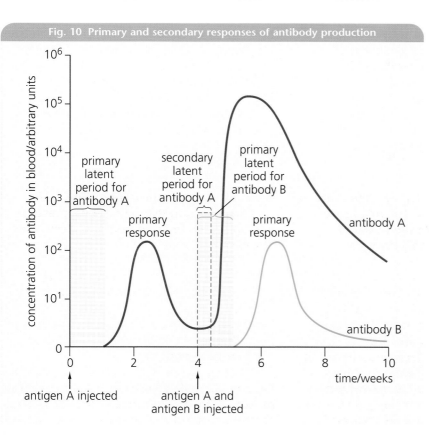

Fig. 10 Primary and secondary responses of antibody production

effective, the immune system must be able to recognise pathogens, and mobilise an effective immune response against them (Fig. 9).

The outside of all cells are covered with chemical substances. **Antigens** are chemical substances that allow the body to distinguish 'self' from 'non-self'. If a pathogen enters the body tissues, the immune system can distinguish its foreign antigens from the body's own antigens. The immune system responds to foreign antigens by making specific substances called **antibodies** and **antitoxins**. Each antibody combines with its specific antigen, and enables one of range of responses that destroy or inactivate the pathogen. Antitoxins neutralise or destroy toxins.

Natural immunity

The first time that an individual is infected with a specific foreign antigen, white blood cells called B lymphocytes are stimulated to produce specific antibodies against these antigens. These specific B lymphocytes divide rapidly by mitosis, so that many cloned cells can produce and release vast quantities of the same antibody into the blood plasma. This is the **primary immune response** (Fig. 10).

Following infection, it takes 3–14 days for the B lymphocytes to start to produce antibody. The period between infection and the onset of antibody production is the **latent period**. After the latent period, the amount of antibody in the blood rises rapidly. The antibody targets the pathogen or its toxins. Infected cells may be destroyed to prevent the spread of the pathogen.

The B lymphocytes that produce the antibody are short-lived. Following destruction of the pathogen, the number of B lymphocytes decreases, causing a reduction in the amount of antibody. However, during the primary immune response, **memory B cells** are produced. These are clones of the B lymphocytes that made the specific antibodies. They remain in the body as a long-term defence against a second or subsequent infection by the bacterium or virus. The individual has now gained **immunity** to the disease-causing organism.

If a second infection by the same pathogen occurs, there is a **secondary immune response**. A much smaller amount of antigen will induce the secondary response because memory cells are already present. When infected, the specific antigen attaches to receptors on the memory cells. This stimulates rapid division of the memory cells, each of which secretes large amounts of the specific antibody. The response is much more rapid than the primary response, and much more antibody is produced. The speed of the secondary response means that the pathogen may be destroyed before the infection takes hold, and symptoms of the disease may be mild or even absent.

7 Using Fig. 10, describe and explain the difference between the primary and secondary response to a pathogen.

Why do we suffer so many colds and bouts of flu?

Viruses have protein spikes on their surface; these are haemagglutinin to attach to host cell membranes, and neuraminidase to destroy host cell membranes when the newly formed virus escapes. The immune system recognises these proteins as antigens in order to mount the immune response, but the proteins also have a significant effect on the ability of the virus to cause disease.

Variation in these proteins is the basis of the different strains of virus. Variation in haemagglutinin and neuraminidase also affects the ability of the virus to cause disease *and* the ability of the immune system to recognise the influenza virus. As antibodies are very specific, they may not combine the antigen of a different strain. By the time antibody is produced against the new strain, the infection has taken hold and produced disease.

Three groups of influenza virus are recognised: type A, type B and type C, but there are many different strains in each group. Type C is most common in young people; type B may cause localised outbreaks of flu, again most commonly in children. Type A is often more virulent, affects all age groups, and may cause flu pandemics.

How do different strains of influenza virus come about?

Mutation in the genetic material of the influenza virus is called **antigenic drift**. It may cause changes in the structure of the viral proteins. This is responsible for most new strains of influenza virus types B and C. In infection caused by type A, infection of host cells by two or more strains of the virus may allow reassortment of genetic material. This is known as **antigenic shift**. It is possible for influenza strains from different host species to combine, and the result may be a completely new and much more virulent strain of influenza. This combination is more likely in rural farming communities where people and animals, such as pigs and poultry, live in close proximity.

Influenza outbreaks occur every year in the UK, and sometimes reach epidemic proportions. In 1989, influenza is thought to have led to the deaths of 25 000 people in the UK. When a new epidemic seems likely, high-risk individuals are identified and contacted, so that they can be offered vaccination. GPs order the amount of vaccine they think they need, often a year in advance, and administer it in preparation for the oncoming winter. Most flu vaccines are cocktails of type A and type B antigens.

8 Explain why someone vaccinated against influenza may still contract the disease.

KEY FACTS

- The primary immune response takes 3–14 days.
- Memory cells produced after the primary response retain the ability to produce specific antibodies for a long time.
- The secondary immune response is very fast because memory cells recognise the antigen.
- Antigenic drift and antigenic shift lead to new strains of virus that the immune system does not recognise.

Artificial immunity

Immunisation is the term for giving individuals **artificial immunity** to a disease so that they are protected without having been infected. In **active artificial immunity**, a weakened or dead form of the pathogen is injected. These preparations are called vaccines. Antigens in the vaccine stimulate the immune system to produce antibodies and memory cells.

Vaccines can be made in several ways. Pathogens can be killed with heat or chemicals to make a **dead vaccine**. The immune system can still recognise the foreign antigens, so antibodies and memory cells are made in response to the vaccine. Havrix vaccine is an example of a dead vaccine against Hepatitis A.

Genetic engineering can be used to produce vaccines. Genes coding for antigens from pathogenic organisms are inserted into non-pathogenic bacteria or yeast cells. The antigens produced by these host cells are extracted and made into vaccines.

Live vaccines (also called **attenuated vaccines**) contain a weakened pathogen that cannot reproduce quickly enough to cause disease. An example is the rubella vaccine. Live vaccines are usually more effective than dead vaccines in stimulating the immune system to produce antibodies and memory cells. However, live vaccines are more likely to produce side-effects.

Tetanus is a disease caused by the bacterium *Clostridium tetani*. The bacteria produce a toxin that causes muscle cramps and can lead to death by heart and breathing failure. Immunisation against tetanus uses a toxoid. When this harmless form of the toxin is injected, B lymphocytes produce antitoxins to the toxoid. Memory cells are also produced. This will protect the individual against the more harmful tetanus toxin produced by *Clostridium tetani*.

Immunity does not always last a lifetime, as memory cells are eventually lost (Table 2). To increase the number of memory cells, 'boosters' are given at intervals.

Vaccines do not guarantee immunity because mutations often produce new strains of the pathogen. For example, vaccination against cholera gives only partial protection against mutant virulent strains of the bacterium *Vibrio cholerae*.

Vaccination schedules for children

Immunisation programmes have helped to reduce mortality in children by reducing the number of children who get the disease. The present vaccination schedule for children is shown in Table 3. Immunisation became

Table 2 Examples of different vaccines			
Disease	**Type of vaccine**	**How given**	**Duration of effect**
diphtheria	toxoid	into muscle	10 years
hepatitis B	genetically engineered antigen	into muscle	many years
influenza	attenuated virus or isolated antigen	into muscle	1–3 years
pertussis	killed bacterial cells	into muscle	many years
rubella	live attenuated strain of virus	subcutaneous	permanent

Table 3 Vaccination schedule for children		
Age	**Immunisations**	**Method**
2 months	*Haemophilus influenzae* type b (Hib)	injection
	diphtheria, tetanus, whooping cough (pertussis)	
	combined injection 1 (DTP 1)	injection
	polio 1	oral
	meningitis C	injection
3 months	DTP 2	injection
	polio 2	oral
	meningitis C	injection
4 months	DTP 3	injection
	polio 3	oral
	meningitis C	injection
12–15 months	measles, mumps, rubella (MMR)	injection
3–5 years	diphtheria, tetanus	injection
	polio booster	oral
10–14 years	BCG (tuberculosis) in high-risk areas unless there is evidence of immunity	injection
11–13 years	rubella (girls)	injection
13–18 years	diphtheria, tetanus booster	injection
	polio booster	oral

Rubella vaccinations have reduced the incidence of this disease in the UK. Serious damage to the nervous system of her fetus is possible if a women contracts rubella during pregnancy.

widely available in the 1940s for diphtheria, in the 1950s for pertussis, and in the 1960s for measles.

Herd immunity

Infectious diseases affect all human populations and are transmitted from a person who has the disease to one who is not immune to the pathogen that causes the disease. The number of cases of the disease increases rapidly in a population with few individuals immune to the pathogen, because sufferers will infect many others before recovery.

The greater the number of people in a population who have immunity, the smaller the chance of the pathogens causing disease in other individuals. It is not necessary to immunise 100% of the population to prevent epidemics of disease. Immunising enough people to prevent spread to the whole population is called the **herd immunity effect**. The required percentage cover is not the same for each disease; this is due to factors such as population density, method of transmission, and the biology of the disease. For instance, 70% cover prevents polio epidemics.

Epidemiologists monitor disease patterns. The possibility of a measles epidemic in 1994–95 was so high that a major publicity campaign was launched to ensure that all school-age children were immunised against the disease. School-age children can become very ill if they get measles; some cases are fatal. The symptoms are a high temperature, a rash, a cough, and sore eyes. Measles can also lead to pneumonia, and may result in blindness, deafness and brain damage. Possible side-effects of the measles vaccine include a mild fever or rash after about a week, but these should not last longer than 3–4 days. There is also a very small risk of brain damage. The dangers of measles greatly outweigh the dangers of the vaccine, and vaccination can be seen to be effective (Table 4). The World Health Organisation plans to eradicate measles by 2010, but must achieve 95% cover in order to do this.

In the 1980s and 1990s, there was a widespread refusal by parents to immunise children against whooping cough because the vaccine was thought to cause serious side-effects including brain damage. There was a subsequent rise in the occurrence of whooping cough; vaccination levels have now gone up again and, following this, the number of cases of the disease have fallen (Fig. 11).

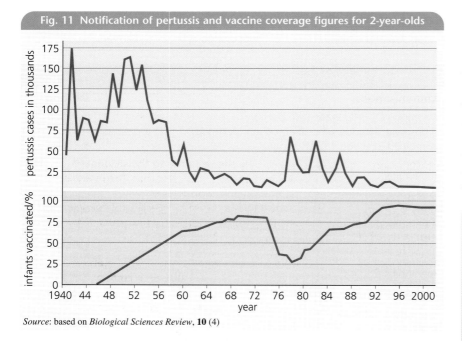

Fig. 11 Notification of pertussis and vaccine coverage figures for 2-year-olds

Source: based on *Biological Sciences Review*, **10** (4)

9a Explain how vaccines give artificial immunity.

b Give two reasons why vaccines may not be 100% effective.

c Which factors may have contributed to a decrease in the number of deaths due to diphtheria, pertussis, and measles?

d Describe the advantages and disadvantages of immunisation programmes.

e Suggest why the percentage cover needed for herd immunity is not the same for all diseases.

Table 4 Notification of measles in England and Wales, 1990 to 1999									
	1990	**1991**	**1992**	**1993**	**1994**	**1995**	**1996**	**1997**	**1998**	**1999**
Measles	13 302	9680	10 268	9612	16 375	7447	5614	3962	3728	2438

Source: Public Health Laboratory Service website, December 2000.

- Immunisation is any method of giving artificial immunity.

- Vaccines produce artificial immunity. A vaccine is a preparation of antigen or pathogen that will stimulate a primary immune response when injected into an individual, but will not cause the disease.

- Children in the UK are usually vaccinated against Hib, diphtheria, tetanus, pertussis, polio, measles, mumps, rubella, and tuberculosis.

- Vaccination and natural exposure to pathogens produce long-term active immunity that relies on the production of antibody and memory cells.

- If enough people are immunised, the incidence of disease is reduced. This is called herd immunity.

1 Smallpox is a disease that has now been eradicated as the result of a vaccination programme organised by the World Health Organisation. The decision to implement the programme of vaccination was partly influenced by the data in the table below.

Year	Total number of people admitted to hospital with smallpox	Percentage mortality of vaccinated people	Percentage mortality of people not vaccinated
1901	1743	9.9	35.8
1902	7916	10.2	33.6
1903	355	2.9	4.5
1904	449	4.8	8.1

a i) Explain what is meant by the term vaccine. (1)
ii) Explain how vaccination can prevent a person developing a disease such as smallpox. (2)

b What evidence is there to show that vaccination is effective against smallpox? (1)

c Apart from the protection offered by the vaccine, suggest **one** other explanation for the difference in percentage mortality between those people who were vaccinated and those who were not. (1)

The table below shows data which relates to the effect of age on the effectiveness of the vaccination. The vaccinated group had each been given a single dose of vaccine in the first year of their lives.

Age/ years	Total number of people admitted to hospital with smallpox	Percentage mortality of vaccinated people	Percentage mortality of people not vaccinated
10–10	1584	1.4	31.9
11–20	1979	1.9	21.8
21–30	3049	5.3	34.5
31–40	2041	13.3	44.4
41–50	995	18.2	55.9
51–60	384	17.7	42.5

i) Describe the effect of age on the effectiveness of the vaccine. (1)
ii) Suggest an explanation for your answer. (1)

AQA/NEAB BY09 March 1998 Q7

2 The graph shows the response of a person to the use of a live vaccine.

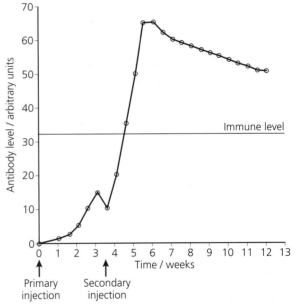

a i) Explain why the response to the secondary injection was much greater than the response to the primary injection. (2)
ii) Suggest why a 'booster' injection of the vaccine may need to be given every few years. (1)

AQA/NEAB BY08 February 1997 Q3a

b The World Health Organisation aims to eradicate poliomyelitis by the year 2005 by increasing infant immunisation in the world to 90%.
i) Suggest how it is possible to eradicate an infectious disease from the world without immunising everyone. (2)
ii) Suggest **two** reasons why the World Health Organisation may not achieve its aim of increasing infant immunisation in the world to 90%. (2)

AQA/NEAB BY09 March 1996 Q5 b, c

8 Lifestyle and health

How we live affects our health. The food we eat, the type or amount of exercise we take, air pollution and smoking, ultraviolet light, our working environment, drinking alcohol, and stress are all factors that affect our health and well-being. To some extent we are in control of most of these, so health promotion campaigns try to encourage us to do the right things to improve our quality of life.

Are there health issues we are certain about? On what is health advice based? What can we do about our own health?

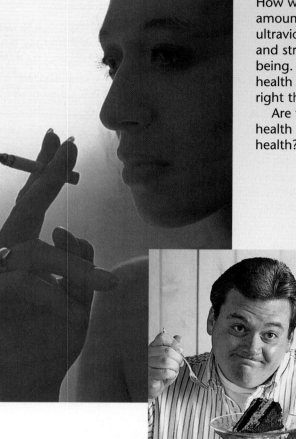

Risks to health sometimes look unattractive, but not always.

A balanced diet

Diet simply means 'the food we eat'. There is a wealth of dietary advice from the government, from TV and magazines, from professional diet writers, and from the food industry itself. The same daily food intake cannot be suitable for all people, because so many factors determine the needs of the individual. For example, age, physical build, and the our usual level of activity significantly affect our nutritional and energy requirements.

The food we eat contains nutrients that are needed for the healthy functioning of the body. A healthy diet provides the right balance of nutrients for the individual concerned. Nutrients can be grouped into six classes:

- carbohydrates;
- lipids;
- proteins;
- vitamins;
- minerals;
- water.

Carbohydrates

Carbohydrates, such as starch and sugar, are the body's main energy sources. Dietary fibre is the name given to complex carbohydrates such as cellulose, which cannot be digested, and so have no energy value. However, insoluble fibre is important in stimulating peristalsis and preventing constipation and bowel disorders such as piles (haemorrhoids) and colon cancer. High-fibre diets affect the absorption of fats, and in turn this can reduce the risk of developing cardiovascular disease.

Lipids

Lipids that are solid at room temperature are called fats, those that are liquid at room temperature are called oils. Lipids are the body's second major dietary energy source and, following digestion, provide essential fatty acids that the human body cannot synthesise. Some lipids, including cholesterol, are essential for cell membranes.

Most lipids from animal sources (e.g. butter and fat in meat) are composed of saturated fats. Plant-derived lipids tend to have a higher proportion of unsaturated fats.

There is a link between the intake of lipids and coronary heart disease (CHD). Government advice, based on reports from health advisory committees, is that we should reduce our total lipid consumption but especially our intake of saturated fats. We are encouraged to consume polyunsaturated fats (see p. 87).

Proteins

Proteins are polymers made up of monomers called amino acids. Proteins are essential for cell growth, repair and replacement. The body uses amino acids to synthesise new proteins in cell division, cell differentiation, muscle and connective tissue, and for enzyme production. Protein is also a structural part of cell membranes.

Nine of the twenty amino acids can only be obtained from the diet. These are called essential amino acids. Others can be synthesised in cells and are known as non-essential amino acids.

Although animal protein contains all the essential amino acids, plant proteins may be lacking in one or more of the essential amino acids.

1a Explain how it is possible to synthesise many different proteins using only 20 different amino acids.

b Suggest why a vegetarian must eat a mixture of plant protein sources.

Minerals

A healthy diet must include an adequate supply of all the minerals we need, so that the body functions properly (Table 1). A deficiency of any particular mineral in the diet may produce recognisable symptoms called a **mineral deficiency disease**. For example, a lack of iron means the body cannot make enough haemoglobin; this leads to a reduction in the number or red blood cells, a condition called anaemia. With insufficient numbers of red blood cells, oxygen transport in the bloodstream is reduced.

2 Explain the effects of a deficiency of the following minerals on a young child:

a iodine;

b calcium.

Vitamins

Vitamins are organic compounds that are needed in very small amounts. Fat-soluble vitamins, such as vitamins A and D, can be stored in adipose tissue. Water-soluble vitamins, such as vitamin C, and are stored to a lesser extent. Vitamins cannot be synthesised, so if any are absent from the diet, a specific **vitamin deficiency disease** result (Table 2, overleaf).

Table 1 Source and function of some dietary minerals			
Mineral	Source	Function	Deficiency disease
calcium	milk, cheese, green vegetables	component of teeth and bone, essential for nerve and muscle function, needed in blood clotting	rickets (bones become soft and bend under body weight)
iron	liver, meat, egg yolk, nuts, legumes	component of haemoglobin and myoglobin, component of respiratory enzymes,	anaemia (low red blood cell count and poor oxygen transport causing lethargy)
iodine	seafood, iodised salt	component of thyroxine	goitre (enlarged thyroid, see Chap. 5)

Table 2 Source and function of some vitamins				
Vitamin	Sources	Function	Deficiency symptoms	Recommended daily allowance
Fat soluble				
A (retinol)	fish liver oil, dairy produce, liver, carrot, spinach	maintenance of healthy membranes, formation of rhodopsin (visual purple)	dry skin, xerophthalmia (dry cornea), night blindness (cannot see in low light conditions)	750 µg
D (calciferol)	milk, butter, fish-liver oil, sunlight on skin	absorption of calcium and phosphorus from gut, calcification of bone	rickets (leg bones soften and bend under body weight)	10 µg
Water soluble				
C (ascorbic acid)	citrus fruit, blackcurrants, tomatoes	collagen formation	scurvy (weak painful joints and blood vessels rupture)	30 µg

3 Explain the importance of fresh fruit and vegetables in the diet.

APPLICATION

Antarctic exploration

In 1914, Shackleton led an expedition aboard the *Endurance* to the Antarctic. The ship became trapped in the ice, then sank. The stranded crew survived on the ice for many months. All were eventually rescued, but were in a weak emaciated condition having survived on a diet of seal, fish and penguin.

1 a Which vitamins did their diet provide?

b Which vitamin was missing from their diet?

Endurance trapped in polar ice.

Water

About 70% of body mass is water. Although the body can survive for several weeks without food, it can only survive for a few days without water. Water is essential because:

- all chemical reactions in the body take place in aqueous solution;
- the surfaces of the lung are moist for gaseous exchange;
- water is the main component of the body's transport systems;
- water can absorb large amounts of heat and is necessary for heat distribution in the blood, and heat loss by the evaporation of sweat.

KEY FACTS

- A balanced diet provides energy, and materials for growth and repair, and will sustain the body functions of the individual. A balanced diet contains carbohydrates, lipids, proteins, vitamins, minerals, and water.

- People vary in their dietary requirements, so the same daily food intake cannot suit everyone.

- Factors that affect dietary needs include age, activity, build, gender, and pregnancy.

- Lack of calcium, iron and iodine each cause a specific mineral deficiency disease.

- Lack of vitamins A, C, and D each cause a specific vitamin deficiency disease.

Diet-related disease

In developed countries, diet-related diseases are usually caused by eating too much. People put on weight and become **obese** if they consume too much energy-rich food and do too little exercise. However, there are other health risks if the nutrient balance is incorrect.

Maintaining the health of the cardiovascular system is very important. People can do much to improve and maintain their cardiovascular health by:

- eating a low-fat diet;
- eating a low-salt diet;
- not smoking;
- reducing stress;
- taking exercise.

Diseases of the heart and circulation are called **cardiovascular disease**; a blocked or burst blood vessel can have serious consequences.

Atherosclerosis

Atherosclerosis is a disease caused by the build-up of fatty deposits, known as **atheroma**, on the inner lining of arteries (Fig. 1). Atheroma can occur in any blood vessel but the usual sites are the aorta, the cerebral arteries in the brain, and the coronary arteries in cardiac muscle.

Fat is transported in the blood as lipoproteins. A diet rich in saturated fat is transported as low density lipoproteins (LDLs), which are more likely to produce atheroma. Cholesterol, although important for making membranes and hormones, is also transported as LDLs. However, poly-unsaturated plant fats are transported as high density lipoproteins (HDLs) and are less likely to be deposited. Studies show that some lipids, such as those found in fish oils, may reduce the deposition of atheroma. Factors contributing atheroma formation include:

- high-fat diet leading to obesity;
- high-salt diet leading to high blood pressure;
- stress leading to high blood pressure;
- smoking;
- age (and menopause in women);
- diabetes;

Atheroma restricts blood flow and increases the chance of a blood clot forming. A **thrombus** is a stationary blood clot and an **embolism** is a mobile clot. A thrombus further restricts the flow of blood, an embolism is carried round the body by the blood, but may eventually lodge somewhere and restrict blood flow.

4 Explain how to reduce the risk of developing atheroma.

Coronary heart disease

Atheroma in coronary arteries is called coronary heart disease (CHD). Narrowing of the coronary arteries may restrict blood flow and starve an area of cardiac muscle of oxygen. This causes a condition called **angina** which results in severe pain usually felt in the

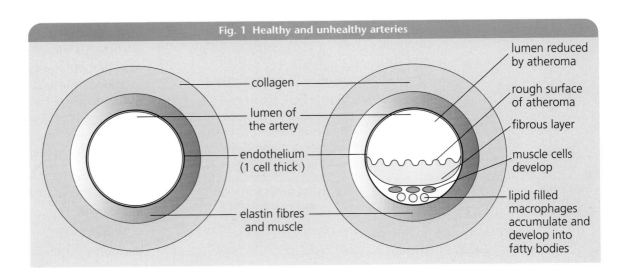

Fig. 1 Healthy and unhealthy arteries

- collagen
- lumen of the artery
- endothelium (1 cell thick)
- elastin fibres and muscle
- lumen reduced by atheroma
- rough surface of atheroma
- fibrous layer
- muscle cells develop
- lipid filled macrophages accumulate and develop into fatty bodies

This light micrograph of a healthy artery has red blood cells in the centre. There is plenty of space for blood flow. In the coloured light micrograph of an unhealthy artery, the build up of material (red and yellow) that has reduced the central space by nearly half, is called atheroma. It is the result of cholesterol being deposited inside the blood vessel. The large mass at the centre (red) is an abnormal blood clot attached to the atheroma. Together they almost block the artery. Atheroma in a coronary artery can lead to a heart attack. Cerebral atheroma can cause cerebral haemorrhage (stroke).

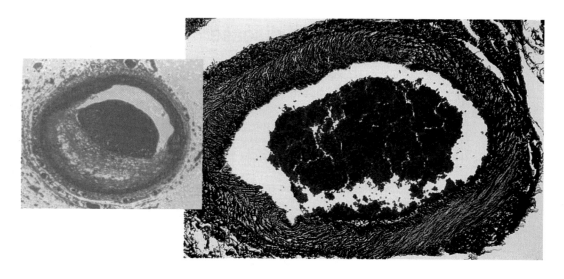

APPLICATION

Benefits of exercise

Exercise is good for you.

People who exercise regularly are better off both physically and psychologically. Regular exercise increases muscle performance; it also affects the circulation and breathing systems of the body. Training increases heart efficiency. The amount of blood pumped by the heart per minute is called **cardiac output** and it varies with activity. As activity increases, cardiac output increases.

cardiac output = heart rate × stroke volume

The difference between the resting output and the maximum the heart can achieve is called the **cardiac reserve**.

Regular exercise leads to a thickening of the cardiac muscle, accompanied by an increase in the size of the heart's chambers. This means that is an increase in the volume of blood pumped by each stroke of the heart. If more blood is pumped with each beat, fewer beats per minute are needed to supply the body with the necessary oxygen and nutrients. So, the average heart rate of a very active person is lower than that of an inactive person. For most people, the resting rate is 65–80 beats per minute, but the resting rate for athletes can be as low as 40–50 beats per minute.

Regular activity also lowers blood pressure. Blood pressure is the force exerted by blood on the walls of the vessels that contain it. Exercise enlarges the blood vessels and improves the blood supply, so reducing blood pressure.

Exercise improves the effectiveness of the breathing mechanism by strengthening the diaphragm and intercostal muscles that control breathing. Gaseous exchange is more effective in both supplying oxygen and removing carbon dioxide. At the cellular level, the metabolic rate is higher, so deposits of atheroma are less likely and less fat is stored in the body, reducing the risk of cardiovascular disease.

1. Which two systems of the body benefit from regular exercise?

2. a A 25-year-old athlete has a resting heart rate of 45 beats per minute, and a cardiac output of 5.25 dm³ min⁻¹. Calculate the stroke volume.

   b An average adult has a stroke volume of 0.06 dm³. Explain the difference between this figure and your answer to 2(a)

3. Study the data in the table on the left.

   a Compare the difference in cardiac output of the athlete and ordinary adult at rest and in vigorous activity.

   b Calculate the cardiac reserve of the average adult and the athlete.

   c Explain the effect of regular exercise on the cardiac reserve.

4. Explain why transport of oxygen to active tissues is more rapid in an athlete due to:

   a the heart;

   b blood vessels;

   c breathing system.

Individual	Cardiac output at rest/dm³ min⁻¹	Cardiac output in vigorous activity/dm³ min⁻¹
average adult	5.0	21.0
athlete	5.25	30.0

chest in men, and in the neck and arms in women.

Blood clots may form in these narrowed blood vessels and block the artery completely, thus depriving the cardiac muscle of its blood supply. Embolisms can form somewhere else in the body and then become lodged a coronary artery. After such a blockage, areas of heart muscle do not function properly and may die. This is called a **myocardial infarction** and can lead to a heart attack when the cardiac muscle fails to contract. This can kill, although many heart attack patients will survive if they are treated quickly.

5a How might a blood clot forming in an artery in the leg eventually cause a heart attack?

b What is this type of blood clot called?

Aneurysm
Artery walls may become weakened due to damage or when elastic tissue is lost. The wall swells into a balloon called an **aneurysm** that may burst and cause serious internal bleeding. Aneurysms occur most frequently in the cerebral arteries and brain. Aneurysms can only be treated by surgery.

Cerebrovascular accident
A thrombus or embolism can block the flow of blood to the brain and deprive the brain of oxygen; this is known as a **cerebrovascular accident** or a stroke. A burst aneurysm will also deprive part of the brain of oxygen.

A stroke can seriously damage or kill cells in brain tissue and, because of the way the brain is organised, the effects of this are often confined to one side of the body. Effects range from slight to severe paralysis, and speech is often affected. Damage may be permanent, but many people do make a gradual recovery. Strokes are more common in older people, but it can happen at any age.

8.3 Environment-related diseases

Environmental factors related to disease include: air pollution and the inhalation of irritant particles, smoking and ultra-violet light.

Diseases of the lung
Air pollution and smoking are responsible for irritation of the lungs which can lead to **chronic bronchitis** and eventually to **emphysema** (Fig. 2). **Carcinogens** in cigarette tars are thought to be responsible for lung cancer in smokers.

Chronic bronchitis
A person is said to have chronic bronchitis if they have had a productive cough (where phlegm is coughed up) that has lasted for at least three months during two successive years. People with chronic bronchitis quickly get out of breath.

Cilia in the bronchial tubes beat rhythmically to produce a constant upward flow of mucus from the bronchi to the back of the throat where it is swallowed. This healthy mucus flow removes trapped dust particles, microbes and other irritants from the bronchi and bronchioles, so they do not reach the alveoli. Smoking and air pollution paralyse the cilia so that the mucus builds up into clumps that are coughed up. The lining of the bronchial tubes becomes irritated and inflamed, with excess thick mucus production. Infections of the breathing

Fig. 2 Development of lung disease

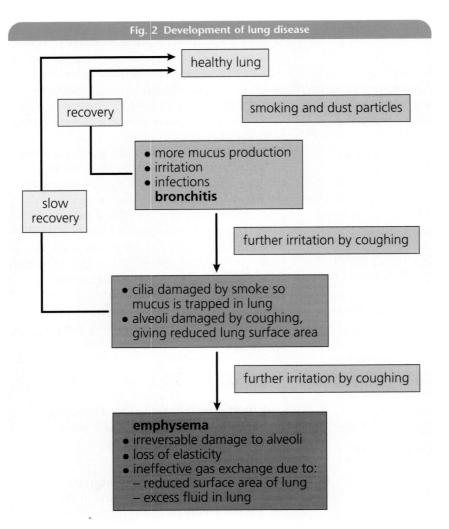

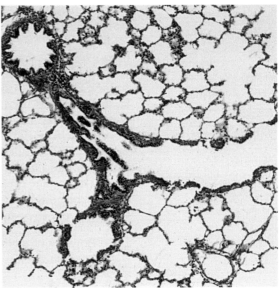

In this healthy lung tissue, the alveoli are folded due to the presence of elastin. They expand during inspiration and contract during expiration to push air out of the lungs.

system such as pneumonia are then much more likely.

Smoking also damages the walls of small bronchioles and alveoli and causes the growth of fibrous tissue around the bronchioles, narrowing the air passageway. This makes it harder to breathe. Bronchitis will get better if a person stops smoking.

6a What damage does smoking do to the bronchi and alveoli?

b What is the effect of this damage?

Emphysema

Smoke and air pollutants irritate the delicate moist surfaces of the lungs. Physical damage by repeated coughing, and loss of elastin from the walls of the alveoli lead to emphysema. In this condition, less surface area is available for the exchange of gases causing breathlessness. Emphysema is common in people who have

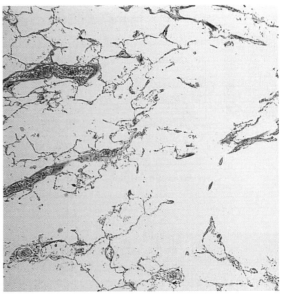

Loss of elastin in emphysema means that the alveoli cannot contract during expiration. Poor ventilation and trapped air causes further damage to airways and alveoli.

smoked for many years, although it is also a linked to a number of occupations, such as coal mining. Emphysema is irreversible. Sufferers are advised to stop smoking and avoid environments with high levels of atmospheric pollutants.

7a Describe how emphysema develops.

b Why does a sufferer of emphysema become breathless easily?

Lung cancer

Carcinogens may cause uncontrolled cell division, producing cancerous tumours. **Benign tumours** stop growing, remain localised, and usually present little health risk. **Malignant tumours** continue to grow and may spread to other parts of the body by a process called **metastasis** (Fig. 3).

Polycyclic aromatic hydrocarbons found in cigarette tar are thought to be the main carcinogens involved in lung cancer in smokers. Lung cancer usually develops in the epithelium lining the bronchioles. A cancer of the epithelium is known as a **carcinoma**. Lung cancer is frequently linked with chronic bronchitis and emphysema. In the early stages, there are often no noticeable symptoms, and the tumour may not show on X-rays. Later, tumours block the air passageways making breathing difficult.

Severe coughing, chest pains, fever, loss of weight and general weakness are all symptoms of the later stages of the disease. Only about 16% of lung carcinomas are diagnosed before metastasis and less than 45% of these patients survive. Less than 1% of patients diagnosed after metastasis survive.

8a Describe how tumours may develop in a smoker.

b Suggest how the growth of tumours in the lung could affect the movement of air through the bronchioles and gaseous exchange.

c What is metastasis?

Skin cancer

Ultraviolet (UV) light is a component of sunlight, and is believed to be the chief cause

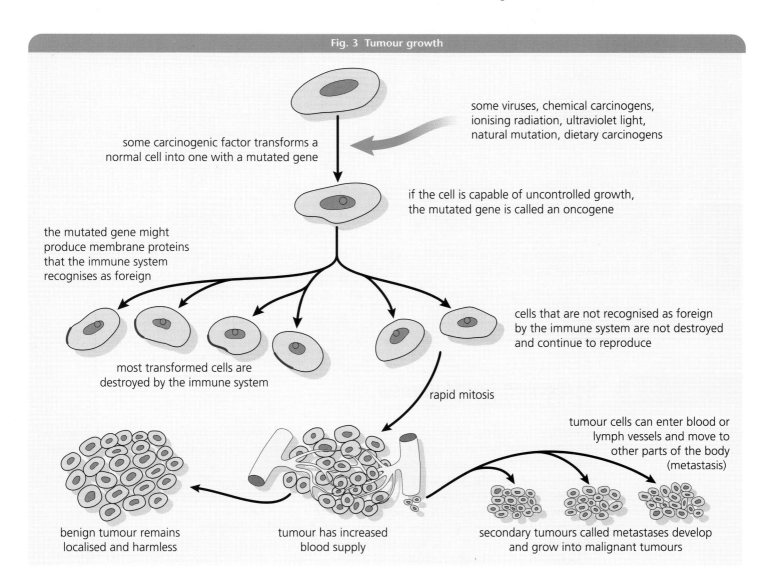

Fig. 3 Tumour growth

some carcinogenic factor transforms a normal cell into one with a mutated gene

some viruses, chemical carcinogens, ionising radiation, ultraviolet light, natural mutation, dietary carcinogens

if the cell is capable of uncontrolled growth, the mutated gene is called an oncogene

the mutated gene might produce membrane proteins that the immune system recognises as foreign

cells that are not recognised as foreign by the immune system are not destroyed and continue to reproduce

most transformed cells are destroyed by the immune system

rapid mitosis

tumour cells can enter blood or lymph vessels and move to other parts of the body (metastasis)

benign tumour remains localised and harmless

tumour has increased blood supply

secondary tumours called metastases develop and grow into malignant tumours of skin cancer. UV light can cause mutations in the skin such that malignant skin tumours arise.

Populations that naturally inhabit sunnier parts of the world tend to have darker skin because of increased amounts of the skin pigment melanin. Melanin gives protection against UV light, so that cells in the basal layers of the epidermis are less easily damaged. Light-coloured skins exposed to the sun gradually increase the amount of melanin (become tanned), but are still more easily damaged by UV light (Table 3).

Studies show that people at risk of skin cancer have light-coloured skin and spend a lot of time in the sun, either living in very sunny climates, or visiting such places for holidays.

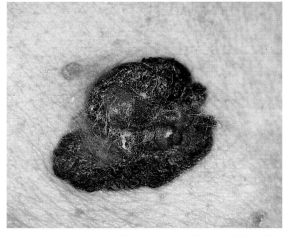

A melanoma begins as a localised growth of skin cells. Moles and freckles are most likely to develop into melanomas.

| Table 3 Skin cancer and place of residence in light-skinned people ||
Location	Number of cases per 100 000 people per year
UK	28
Africa (Cape)	133
Texas	168
Queensland (Australia)	265

Protective creams with 'sun screens' reduce the effect of UV light, but in areas of strong sunlight, extra care should be taken. Treatment of skin cancer before any significant metastasis has a success rate of over 90%, but the success rate of treatment after metastasis is less than 15%.

9a How is skin cancer commonly caused?

b Suggest two pieces of advice to reduce the chances of developing skin cancer.

KEY FACTS

- Lung diseases are caused by smoking and air pollution.

- Bronchitis is characterised by a persistent cough and excessive mucus production.

- Break down of alveoli in the lung results in emphysema.

- Lung diseases impair gaseous exchange and reduce breathing efficiency.

- Carcinogens produce tumours in the lungs.

- Malignant cancers spread by metastasis; benign tumours do not develop further.

- Skin cancer is caused by UV light.

8.4 Screening programmes

Screening is looking for signs of disease when the patient may not be aware of any symptoms; it enables early diagnosis. Sometimes screening is random, for example, a doctor may decide to check the blood pressure of all adult patients who come to the surgery. Other screening programmes may be systematic, for example, the screening of schoolchildren for hearing or sight problems.

Some screening is of self-selecting groups, for example, women who go for cervical smear tests.

There is a wide range of screening and diagnostic techniques (Table 4).

Genetic analysis
Genetic disorders may affect a child from birth, or may develop later in life (Table 5).

Table 4 Diagnostic and screening techniques		
Chemical	**Biological**	**Physical**
biochemical tests (e.g. urine tests)	biopsies (on tissue samples)	X-rays
immunological tests (for antibodies)	cytological examination culturing microorganisms genetic analysis eye tests hearing tests	ultrasound endoscopy blood pressure

Table 5 Some genetic conditions	
Condition	**Genetics**
cystic fibrosis	autosomal recessive
haemophilia	X-linked recessive
Duchenne muscular dystrophy	X-linked recessive
sickle cell disease	autosomal co-dominant
thalassaemia	autosomal co-dominant

Tests during pregnancy

Many genetic conditions can be detected during pregnancy by chorionic villus sampling or amniocentesis. Chorionic villus sampling is performed 10 weeks after conception and has a 1–2% risk of miscarriage. A sample of the placenta is taken via the cervix (Fig. 4). The samples are examined for chromosome abnormalities or by DNA tests.

Amniocentesis is performed 15–20 weeks after conception and has 0.5–1% risk of miscarriage. A sample of amniotic fluid is collected using a fine needle passed through the abdominal wall into the amniotic cavity and guided by ultrasound (Fig. 5). Cells from the sample are cultured for three weeks before examination or genetic analysis.

Down's syndrome is caused by an extra chromosome number 21 and is more common in children born to older mothers. This can be identified in dividing cells when chromosomes are visible during mitosis. Individuals with Down's syndrome have

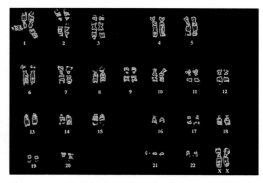

Chromosomes in a dividing cell can be photographed and then sorted into homologous pairs. These arrangements are called karyograms. This is the karyogram of a Down's syndrome girl.

changed facial features, short stature, and learning difficulties. With care and support, people with this syndrome can become self-sufficient and lead an active life.

Tests before pregnancy

It is possible to detect the genes for some diseases in people before they become

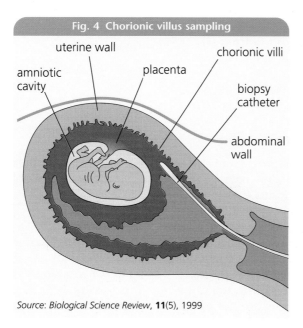

Fig. 4 Chorionic villus sampling

uterine wall
amniotic cavity
placenta
chorionic villi
biopsy catheter
abdominal wall

Source: Biological Science Review, **11**(5), 1999

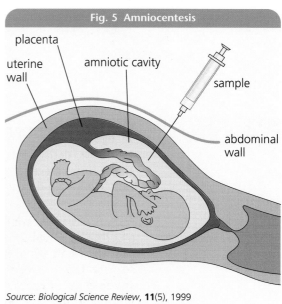

Fig. 5 Amniocentesis

placenta
uterine wall
amniotic cavity
sample
abdominal wall

Source: Biological Science Review, **11**(5), 1999

parents. If a man and woman both carry the recessive cystic fibrosis allele, there is a chance that their child could inherit the condition. A test on a saliva sample can determine whether or not a person is carrying a recessive cystic fibrosis allele.

Professional advice is available for those who wish to take advantage of it. This help is called genetic counselling and is very important to those who have to make difficult decisions about the chances of passing on genetic disorders, and whether or not to continue or start a pregnancy.

Ethical issues

Confidentiality about the use of information gained by genetic techniques concerns many people and raises many questions. Who has a right to the information? Should parents make the decision for an unborn child with an inherited condition? Should insurance companies know about people with inherited conditions?

10

a Describe one advantage and one disadvantage of chorionic villus sampling.

b How can you tell if a fetus has Down's syndrome?

11

a One in twenty-one people carry the recessive cystic fibrosis allele. What is the probability of two carriers marrying?

b A man and woman both carry the recessive cystic fibrosis allele. Using a genetic diagram, show the chance of their child (i) inheriting cystic fibrosis; (ii) becoming a carrier for cystic fibrosis.

12 What ethical issues arise from using genetic screening?

X-rays

In X-rays, dense objects appear white and soft tissue looks black. X-rays show damage to bones and can detect abnormalities in soft tissue. For example, certain types of cancer can be detected this way. As X-rays are a form of ionising radiation and can damage DNA, care has to be taken as too many X-rays could be harmful, and could even cause cancer. Women's ovaries are always protected from X-rays, and pregnant women are not usually

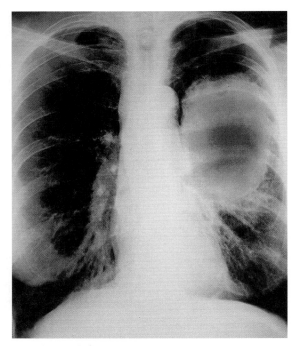

In this false-colour X-ray, the bony structures are shown in yellow and healthy soft tissue is purple. The large red mass is a carcinoma in the left lung (right-hand side of image) of a smoker.

X-rayed. The **radiographer** always works from behind a protective screen to avoid over-exposure to the radiation.

13

a What sort of problems can be diagnosed with X-rays?

b Suggest why X-rays are not used for regular routine screening in smokers.

Ultrasound

Ultrasound scanning directs very high frequency sound waves into the area being investigated, and converts the reflected sound into visual images. Unlike X-rays, ultrasound does not damage DNA.

14 Why is ultrasound used to monitor fetal development but X-rays are not?

Endoscopy

An endoscope is an instrument for looking inside the body. It can be used, for example, to look for tumours or ulcers in the stomach. It is usually inserted through a natural opening, but small incisions may be made to examine other parts of the body. Short rigid endoscopes are used for keyhole surgery on joints or within the abdomen. Longer flexible endoscopes up to a metre long can be manipulated within the body for diagnosis and treatment (Fig. 6).

15 Explain how a flexible endoscope could be used to examine a lung, and how a sample of tissue could be collected for examination.

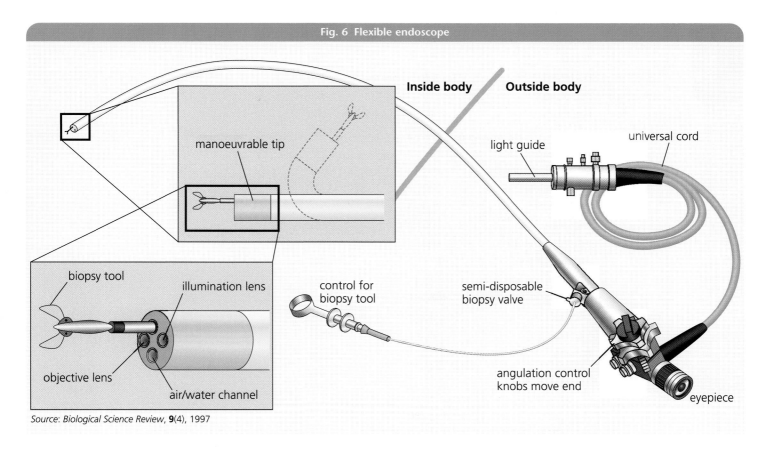

Fig. 6 Flexible endoscope

Inside body　　Outside body

manoeuvrable tip

light guide

universal cord

biopsy tool

illumination lens

control for biopsy tool

semi-disposable biopsy valve

objective lens

air/water channel

angulation control knobs move end

eyepiece

Source: Biological Science Review, **9**(4), 1997

KEY FACTS

■ Screening enables early diagnosis and treatment, so increasing the patient's chance of recovery.

■ Screening techniques must be reliable and safe, and allow effective treatment to be carried out.

■ Genetic screening can show whether or not a fetus has an inherited condition. It also enables prospective parents to discuss the risk of passing on genetic conditions to their offspring.

■ Genetic screening raises many ethical issues.

1 The table shows an analysis of the food eaten in one day by a 15-year-old girl.

Meal	Food	Mass/g	Energy content/kJ	Carbohydrate/g	Fat/g
breakfast	cornflakes	25	390	21.3	0.4
	milk	100	270	5.6	3.9
	egg	60	370	0	6.5
	bacon	40	560	0	12.6
	bread	30	290	14.5	0.5
	butter	8	240	0	6.6
lunch	fish fingers	110	820	17.7	8.3
	ketchup	15	60	3.6	0
	chips	130	1280	44.2	13.3
	fruit yoghurt	150	570	27.0	1.0
supper	steak and kidney pie	160	1830	35.5	27.4
	chips	180	1770	61.2	18.4
	baked beans	150	520	22.7	0.9
	chocolate cake	100	2100	53.1	30.9
drinks	instant coffee (5 cups)	20	80	2.2	0
	milk (in drinks)	125	340	5.8	4.9
	sugar (in coffee)	25	420	25.0	0
	canned drinks	600	1000	60.0	0
snacks	chocolate	50	1110	29.7	15.2
	biscuits	60	1160	44.9	10.0
	banana	130	420	25.0	0
	orange	100	150	8.5	0
totals		2368	15750	507.5	160.8

a The recommended energy intake per day for a 15-year-old girl is 9.6 MJ.
  i) By how much did the girl's diet exceed this? (1)
  ii) Describe the likely long-term effects on her blood vessels if she continued to exceed the recommended intake by this amount. (3)

b Some nutritionists consider it more healthy to eat fats derived from plants than from animals.
  i) Why is eating animal fats considered to be less healthy than eating plant fats? (2)
  ii) Calculate the percentage of fat in the girl's breakfast that was obtained from animal foods. Show your working. (2)
  iii) Explain why it would be difficult to calculate the percentage of animal fat in the girl's supper. (1)

AQA/NEAB BY08 February 1996 Q1

2
a Tuberculosis is a bacterial infection of the lungs that may cause extensive damage to lung tissue. Based on this information suggest why:
  i) the use of X-rays is a suitable technique to detect tuberculosis; (1)
  ii) mass screening was successful in reducing the incidence of tuberculosis during the 1950s and 60s in the UK. (1)

b i) What is an endoscope?
  ii) How may an endoscope be used as a way of detecting damage to lung tissue? (2)

c A recent development in screening for genetic disease is the use of single strand DNA called gene probes. These are radioactively labelled strands of DNA carrying a complementary sequence to the mutant gene responsible for the disease. The main stages in using a gene probe are shown in the flow diagram.

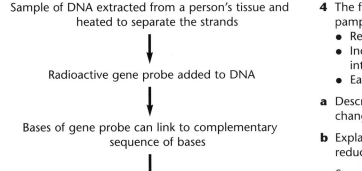

Sample of DNA extracted from a person's tissue and heated to separate the strands

↓

Radioactive gene probe added to DNA

↓

Bases of gene probe can link to complementary sequence of bases

↓

Excess probe washed away

↓

Sample tested for radioactivity

Using this information, explain how a gene probe can be used to detect the presence of a mutant gene. (3)

AQA/NEAB BY08 February 1997 Q4

4 The following recommendations are taken from a pamphlet on healthy eating.
  ● Reduce fat intake from all sources.
  ● Increase dietary fibre (non-soluble carbohydrate) intake.
  ● Eat less salt.

a Describe and explain the likely effect on health of changing to a lower fat and higher fibre diet. (6)

b Explain why the amount of salt eaten should be reduced. (3)

c Suggest why medical opinion favours a combination of exercise and diet, rather than diet alone, to lose weight. (3)

AQA/NEAB BY08 February 1997 Q7

3 The diagram shows a section through the wall of a bronchus of a healthy person.

a i) Give **two** ways in which the bronchus shown in this diagram differs from a bronchus in a person suffering from chronic bronchitis. (2)

  ii) Explain how each of these differences contributes towards the symptoms of chronic bronchitis. (2)

b Suggest how the link between cigarette smoking and the incidence of chronic bronchitis could be investigated. (3)

AQA/NEAB BY08 June 1999 Q3

cilia

goblet cell

epithelium

mucus secreting gland

smooth muscle

cartilage

cartilage cells

connective tissue

Source: adapted from Freeman and Bracegridle, *Atlas of Histology*, Heinemann Educational Publishers

Answers to in-text questions and questions in application boxes are given below. Answers to examination questions are not included.

Chapter 1

1

a Young canada geese follow their parents and learn traditional migration routes and winter feeding grounds. Geese reared in captivity have no knowledge of migration routes, though they might develop new ones given time.

b Migration behaviour in blackcaps is innate. Birds reared in captivity will therefore show the correct migration pattern.

2

a If the flatworms used a taxis, they would use light receptors to compare the light intensity on each side, and move to the darker side. If they emerged into the light area they would immediately turn back.

b If the flatworms used a kinesis, they would move in straight lines until they found themselves in a dark enough environment. They would then turn in tight circles and/or slow down their rate of movement. (This is what flatworms actually do.)

Application boxes
Habituation in babies

1

a The startle reaction is a quick, innate response to the sound stimulus.

b Learning occurs when there is a change in the response to a stimulus. Here, the response ceases, so we can say the response has changed.

c Habituation occurs to only a specific note (or one very close to it).

d After habituation to one note, a note of slightly different pitch can be played to the baby. If the baby shows a startle response, it can tell the difference between the two notes.

e Babies that do not habituate are repeatedly disturbed when feeding. However, not responding to a *new* sound might mean not responding to danger.

Saving the takahe

1

a The young birds would not imprint on takahes. They might not recognise their own species when released.

b The takahes would already be imprinted so it would matter less at this stage.

2

a Takahes generalise the image of the model to similar looking objects, so a rough model is sufficient.

b Breeding between close relatives increases the chances of harmful recessive alleles becoming homozygous, and therefore being expressed.

Chapter 2

1 Stickleback models can be made with different colour patterns and shapes. These should be placed near a male stickleback in a standardised way and the responses of the male fish recorded. The experiment needs to be repeated many times with many fish to determine an overall pattern. When the scientist Tinbergen did this in the 1950s, he discovered that male sticklebacks attack red objects regardless of their shape, but show little response to brown models, even if they are fish-shaped.

2

a Courtship feeding helps to overcome the fear of the red breast's aggressive signal and also allows the female to select a mate who is good at collecting food.

b The female needs extra food to form eggs and will continue to need food while she is confined to the nest.

3 Similarities: a hormone and a pheromone are both chemical signals that influence physiology or behaviour; they are made by animals.
Differences: a hormone travels in the bloodstream to other parts of the same animal; a pheromone travels in the air or water to other members of the species.

4 Advantages: she is able to attract males and be mated when she is most likely to conceive; she may attract several males and be able to choose the fittest.
Disadvantages: she may be deserted by her mate when she is no longer fertile. By disguising the state of oestrus, a female obliges a male to stay with his mate in order to be confident of his paternity; a longer partnership ensures that the male remains to help protect and rear his offspring.

5

a More chicks were fledged per pair in the better quality nest sites.

b Shags cannot defend a feeding area because they feed out to sea on shoals of fish which are continuously moving.

Application boxes
Feeding herring gull chicks

1 Age of chick, number of trials with each chick, distance and orientation of model, temperature, how recently taken from nest, surroundings etc.

2

a A single chick may not be typical of all chicks.

b Chicks may have individual preferences for beak colour etc., but there is a need to get an overall picture and to take the average of all the results. A chick may become tired, habituated to the stimulus or give up because it gets no reinforcement. The chick might suffer from exposure, fright, or lack of food if used for too long.

3 Experiment 1: Chicks showed a preference for red bills over other colours.

Experiment 2: Chicks pecked most at red spots rather than spots of other colours.

Experiment 3: Chicks pecked most at spots with the greatest contrast between the background tone and the spot (i.e. a white spot on a dark beak or a dark spot on a white beak).

Some models were pecked at more than the natural coloured one (red spot on yellow beak); an all-red beak and a white spot on a dark beak were

Chapter 7

1

a A pathogen is a microorganism that causes disease.

b Infection is the presence of viable microorganisms in the host's body. Disease is when the host shows the recognisable signs and symptoms caused by the presence of the organism.

c In carriers, a pathogen may be present in the body tissues, but the individual does not show signs or symptoms.

d Percentage of TB carriers who develop the disease =

$$\frac{20\,000\,000}{1\,900\,000\,000} \times 100\% = 1.05\%$$

2 Sterile clothes, the use of face masks for hospital staff, and filtered air (maintained above atmospheric pressure) all reduce the risk of air-borne infection in operating theatres.

3 Settling tanks and filtration remove microorganisms and organic debris, which could otherwise act as a food source for microorganisms. Chlorine kills microorganisms in the water.

4

a High temperatures denature enzymes and kill microorganisms in the food.

b Washing hands removes microorganisms living on the skin surface, and microorganisms that may have come from handling other foods.

c Cool temperatures reduce the growth rate of microorganisms. Provided storage at cool temperature is for only a short time, the microorganisms cannot reproduce to potentially dangerous numbers.

d Pathogenic microorganisms could be transferred from uncooked to cooked foods if they were kept together.

5 Washing hands regularly to remove microorganisms; condoms reduce the chances of transmission by sexual contact; use of new or sterile needles and medical equipment to reduce transfer from one patient to another; screening of blood and blood products to prevent transmission by transfusions.

6

a Droplet infection:
physical barriers – skin, mucus in airways;
chemical barriers – lysozyme in tears and saliva.

b Water-borne and food-borne infection:
physical barriers – mucus in gut;
chemical barriers – lysozyme in saliva, hydrochloric acid in stomach.

c Contact infection:
physical barriers – skin, mucus;
chemical barriers –lysozyme in tears and saliva.

7 The primary response has a longer latent period and the concentration of antibody reaches lower levels (just over 10^2 arbitrary units) than in the secondary response (just over 10^5 arbitrary units). The rate of antibody production is greater in the secondary response.
During the primary response, it takes 3–14 days for clones of B-lymphocytes to produce antibody. In the secondary response, the foreign antigens are recognised by memory cells, which can be cloned quickly to produce larger amounts of antibody more rapidly.

8 Vaccination provides immunity to a particular type of flu. The structure of the protein spikes of the influenza virus may change due to antigenic drift and antigenic shift. These new strains of virus may not be recognised by the immune system, and so cause disease.

9

a Antigens in the vaccine stimulate the immune system to produce specific antibodies and memory cells. This gives protection against the pathogen from which the antigens were derived.

b Reasons for vaccines not being 100% effective include:
- antibody levels fall after vaccination;
- memory cells are gradually lost;
- variation occurs in the pathogen so some strains may not be recognised by the immune system.

c Although immunisation programmes have contributed to the eradication of all three diseases (particularly diphtheria), the main reason for fewer deaths is improvement in social conditions. Better nutrition, less overcrowded housing, and improved personal hygiene are significant factors.

d Advantages: decreased death and suffering; reduced treatment provision by the National Health Service; less loss of school or work time. Disadvantages: risk of vaccine side-effects; immunisation does not guarantee immunity; cost of vaccines and administering vaccination programmes.

e The pathogens that cause disease are transmitted in different ways; some diseases are more infectious than others; the duration of illness and type of disease may increase the chances of transmission; diseases may be transmitted by symptomless carriers; individuals may have natural immunity after suffering from the disease.

Application boxes
Flu

1

a Winter.

b People spend more time indoors, and ventilation tends to be poor. The air is damp, which assists droplet infection.

2

a
- Avoid crowded locations
- Ventilate rooms
- Avoid infected people
- Wash hands regularly
- Do not share cutlery, crockery, or drinking vessels
- Have a flu vaccination

b Children: School-based health promotion campaigns.
Elderly people: doctors, heath centres, and pharmacies could target high-risk weaker and older patients; government could provide an information leaflet when pension is collected.

Chapter 8

1

a Both the order and composition of the amino acids determine the structure, and hence the properties, of a protein. Thus, millions of proteins can be synthesised from just 20 amino acids.

b Plants contain proportionally less proteins than animals, and essential amino acids are not present in some plant products. Vegetarians must ensure they consume enough plant protein and choose plant products that together contain all essential amino acids.

2

a Lack of iodine would lead to poor growth and poor mental development (cretinism) because iodine is an essential component in the hormone thyroxine.

b Lack of calcium would lead to weak bones that could bend under the weight of the body (rickets) as calcium is required for bone growth.

3 Fresh fruit and vegetables are important sources of vitamin C, calcium and dietary fibre. However, cooking reduces the amount of vitamin C in fruit and vegetables.

4 A diet low in saturated fat decreases LDLs in the blood, thus reducing atheroma formation. Reducing stress, not smoking, and lowering salt intake all reduce blood pressure, a significant factor in atheroma formation. Regular exercise improves peripheral circulation, increases metabolic rate, makes obesity less likely and so reduces the risk of atheroma formation.

5

a The blood clot from the leg may be carried into the coronary artery of the heart. Reduced blood flow in the cardiac tissue prevents the tissue functioning, and a heart attack is the result.

b A mobile blood clot is called an embolism

6

a Chemicals in the cigarette smoke paralyse and destroy cilia in the bronchial tubes. The cilia normally clear the tubes of mucus, which carries air-borne particles such as dust, bacteria and fungal spores. Inflammation of the tubes may occur. Smaller airways become narrower due to the growth of fibrous tissue, and alveoli may become blocked.

b Trapped microorganisms may cause infections. Inflammation and mucus may block airways preventing gaseous exchange. Persistent coughing damages the alveoli and reduces lung surface area. All contribute to reduced gaseous exchange and difficulty in breathing.

7

a Tobacco smoke irritates the cells lining the bronchi and bronchioles. Increased mucus production and reduced numbers of cilia lead to persistent coughing. The result of coughing is damage to the alveoli.

b Damage to the alveoli by coughing reduces the surface area of the lung, and loss of elastin reduces the ability of the lung to expel air during expiration. Reduced gaseous exchange fails to supply sufficient oxygen and remove waste carbon dioxide from the blood. Breathing rate increases causing breathlessness, even at rest.

8

a Carcinogens in tobacco smoke are carried into the lung, and cause uncontrolled growth of the epithelial cells in the lung. Tumours are produced.

b Tumours block airways and thus prevent air flow. Tumours replace alveoli reducing the surface area for gaseous exchange.

c Metastasis is the movement of malignant cells from tumours to other parts of the body, where they subsequently form secondary tumours.

9

a UV light in sunlight causes damage to the DNA in the skin cells. This causes skin cancer as a result of uncontrolled growth of cells in the basal layers of the epidermis.

b **(i)** Reduce the time and frequency of exposure to UV light.
  (ii) Use a sun screen to reduce the effect of harmful UV light.

10

a Advantage: the technique permits the diagnosis of genetic disorders before the birth of a child.
Disadvantage: the technique carries a 1–2% risk of miscarriage.

b Fetal cells can be collected by amniocentesis and used to construct a karyogram. The presence of an extra chromosome number 21 (three instead of two) shows the fetus has a Down's syndrome karotype.

11

a $21 \times 21 = 441$; so the probability of two carriers marrying is 1 in 441.

b

	Man	Woman
	Cc	Cc
Gametes	Ⓒ ⓒ	Ⓒ ⓒ
Children	CC Cc	Cc cc
	Normal carrier	carrier cystic fibrosis sufferer

C = normal *cftr*
c = mutant *cftr*

  (i) There is a 1 in 4 (25%) chance that the child will have cystic fibrosis.
  (ii) There is a 1 in 2 (50%) chance that the child will be a carrier of the cystic fibrosis allele.

12 Ethical issues arise from using genetic screening include:
  • termination of pregnancy;
  • future problems caused by disease;
  • letting potential sufferers know;
  • who should have the information.

13

a X-rays can be used to diagnose bone injuries, and tumours in soft tissues.

b X-rays increase the risk of causing cancers by damaging DNA.

14 Unlike X-rays, ultrasound does not damage DNA, and allows the development of the fetus to be monitored.

15 The endoscope tube is inserted via the mouth. It would be manipulated to the area of the lung under examination. A biopsy tool would allow an operator using the eyepiece to remove a sample of tissue.

Application boxes
Antarctic exploration

1

a Vitamins A and D would have been provided by the crew's diet (and also vitamin B if they ate the livers of the animals).

b The vitamin missing from the crew's diet was vitamin C.

Benefits of exercise

1 The cardiovascular system (heart and blood vessels) and the breathing system benefit from regular exercise.

2

a Stroke volume = 5.25/45 = 0.117 dm^3

b An average adult has a lower stroke volume than an athlete because regular exercise increases the size of the heart chambers thus increasing the stroke volume.

3

a The athlete has a slightly higher cardiac output at rest (0.25 dm^3 min^{-1} higher) and a much higher output during exercise (9 dm^3 min^{-1} higher).

b Cardiac reserve of the average adult
= 21 − 5 = 16 dm^3 min^{-1}
Cardiac reserve of the athlete
= 30 − 5.25 = 24.75 dm^3 min^{-1}

c Regular exercise increases the cardiac reserve as the heart becomes larger, the cardiac muscle is stronger, and the peripheral resistance to blood flow is reduced due to development of the peripheral circulation in the muscles. These all contribute to an increased cardiac reserve.

4

a Greater cardiac output increases blood flow to active tissues.

b Increased capillary networks in the muscles improve gaseous exchange.

c Improved capillary networks in the lungs, increased lung volume, and better ventilation improve gaseous exchange in the lungs.

accessory gland
One of the glands close to the male reproductive organs. Accessory glands add materials to the semen.

acrosome reaction
Release of enzymes stored in the acrosome.

acrosome
A membrane-bound sack in the sperm head, containing enzymes.

active artificial immunity
The immunity that arises when an individual actively produces antibody in response to being given an antigen preparation.

adolescence
A period of physical growth and sexual development, that divides childhood from adulthood.

ageing
Changes in the structure and function of an individual that occur with time.

allometric growth
The term to describe the different growth rates of different organs and the entire organism.

aneurysm
A balloon-like swelling in an artery wall. It occurs when artery walls are weakened due to damage or loss of elastic tissue.

angina
Pain caused by poor blood flow in the cardiac tissue.

antibody (immunoglobulin)
Protein molecules produced by B lymphocytes to combat infection and provide immunity.

antigen (immunogen)
A substance that stimulates an immune response.

antigenic drift
Mutations that cause variation in the antigens of a pathogenic organism.

antigenic shift
Changes in antigens from the combination of genetic material from two or more strains

antitoxin
A substance that neutralises the toxins produced by pathogens.

artificial immunity
Immunity produced by vaccination.

associate
To link one stimulus with another in the process of learning.

atheroma
Fatty deposits that accumulate on the inner lining (endothelium) of an artery.

autoimmunity
A condition in which the immune system fails to recognise body cells as 'self', and then attacks them.

average life expectancy
The age at which 50% of a population have died (and 50% remain alive).

benign tumour
A tumour of cancerous cells that stops growing and remains localised.

biological control
Using a predator or naturally occurring attractant to control a pest, rather than using synthetic pesticides.

birth control
Ways of preventing unwanted children, such as contraception and abortion.

birth rate
The number of births per thousand of the population per year.

blastocyst
The stage of an embryo when it is a hollow ball of cells.

capacitation
Changes occurring in the sperm, after ejaculation, making them capable of fertilising an oocyte (egg).

carcinogen
A substance that causes cancer.

carcinoma
A cancer of epithelium tissue.

cardiac output
The volume of blood pumped per ventricle per minute measured in $dm^{-3}\ min^{-1}$.

cardiac reserve
The difference between resting output and the maximum output the heart can achieve.

cardiovascular disease
Disease of the heart or circulation system.

carrier
An individual who is carrying a pathogen but does not show signs or symptoms of a disease.

cerebrovascular accident
A blockage in the blood supply to the brain, caused by a thrombus or embolism and resulting in damage to brain cells.

chorionic villus (plural, villi)
Finger-like extensions of the placenta containing capillaries, surrounded by the chorion membrane.

chronic bronchitis
A persistent productive cough, where phlegm is coughed up, lasting for a period of at least three months, occurring during two successive years.

chronological age
An individual's age from birth.

classical conditioning
The conditioning of a reflex action so that an animal performs the reflex in response to a new stimulus.

conception
The implantation of an embryo in the uterus wall.

conditioning
Associative learning (conditioning is divided into classical conditioning and operant conditioning).

contraception
Methods of birth control by preventing conception.

corpus luteum
The 'yellow body' made of cells derived from the ruptured Graafian follicle in the ovary, making hormones such as progesterone.

cortical granule
A membrane-bound vesicle that releases enzymes after a sperm has penetrated an oocyte. These enzymes cause the cortical reaction.

cortical reaction
Changes occurring to the zona pellucida making it impenetrable by further sperm.

counter-current
The movement of fluids in opposite directions at each side of an exchange surface, allowing more rapid diffusion.

cross-sectional study
A sample of people of differing ages who are studied at the same time. This allows individuals of differing ages to be compared simultaneously.

dead vaccine
A vaccine containing killed pathogens; the organisms cannot cause disease but do stimulate an immune response.

death rate
The number of deaths per thousand of the population per year.

demographic transition model
A model to explain the changes in birth rate, death rate and total population in response to social and economic factors.

disease
The dysfunction of the body's normal functions.

dominance hierarchy
A social ranking system in animals. Position in the hierarchy determines which animal takes precedence over others in the group for food, mates or space.

doubling time
The time, in years, for a population to double in size (based on its current growth).

ecological niche
An organism's place and function in its environment.

embolism
A mobile blood clot.

embryo
A group of cells developing from a zygote, before the development of organs.

emphysema
A lung disease in which the alveoli collapse. This reduces the lung surface area and the effectiveness of gaseous exchange. Sufferers are breathless even at rest.

endometrium
The inner lining of the uterus, which has a rich blood supply.

environmental factor
An external physical, chemical or social factor that affects living organisms.

epidemic
A sudden and widespread outbreak of a disease.

extended period of dependency
The long childhood that enables primate offspring to learn and develop social skills; human offspring are dependent on their parents for very long a period of time.

feedback effect
The regulation of the concentration of a substance in the body in which the substance's concentration influences its own production or release.

female pronucleus
The haploid set of chromosomes produced in an oocyte, which fuses with the male pronucleus at fertilisation.

fertilisation membrane
The impenetrable barrier produced by the zona pellucida after fertilisation, as a result of the cortical reaction, preventing the entry of further sperm to the oocyte.

fertilisation
The fusion of a male and female pronucleus to produce a zygote.

fertility treatment
Treatment to help childless couples to have children.

first cleavage
The first mitotic division of the zygote.

follicle stimulating hormone (FSH)
A hormone produced by the anterior pituitary gland, causing the maturation of follicles and other changes during the menstrual cycle.

follicle
A group of cells surrounding a developing oocyte in the ovary, providing nutrients and making hormones.

gestation period
The time from conception to birth.

glycoprotein
A substance made from carbohydrate joined to protein. Glycoproteins occur on the surfaces of cells.

gonadotrophic hormone
A hormone produced by the pituitary gland that stimulates the ovaries or testes.

Graafian follicle
A mature follicle containing a secondary oocyte and swollen with fluid. A mature Graafian follicle is able to ovulate.

granulosa cell
A cell of the follicle surrounding the secondary oocyte. Some granulosa cells remain around the oocyte after ovulation.

growth curve
A graphical representation in which a growth measurement is plotted against time.

growth rate
Change in size per unit time.

habituation
Ceasing to respond to a frequently repeated stimulus. This is a simple type of learning.

herd immunity effect
The effect of immunising a sufficiently large number of people to protect an entire population from the spread of a particular disease.

hormone
A chemical substance, released from an endocrine gland, which travels in the bloodstream in small concentrations. The hormone travels throughout the body, and changes the metabolism or physiology of one or more organs, called the target organs.

human chorionic gonadotrophin (hCG)
A hormone released by the implanted embryo and which stimulates the corpus luteum of the mother's ovaries.

hyperthyroidism
The condition in which the thyroid gland is over-active, and produces excess thyroxine.

hypothalamus
The region of the forebrain that controls activities such as temperature regulation, osmoregulation and hormone release. It is connected to the pituitary gland and thereby links the nervous system to the endocrine system.

hypothyroidism
The condition in which the thyroid gland is under-active, and produces too little thyroxine.

immune system
The system that resists the effects of pathogenic organisms that have entered the body.

immunisation
The process of artificially creating immunity in an individual.

immunity
The ability of an organism to resist infection.

imprinting
A type of learning in which birds that leave the nest shortly after hatching follow the first moving object they see. The young birds then continue to follow that object and no other.

incubation period
The period of time between infection and the development of signs and symptoms.

individual space
The space maintained between individuals of group-living species.

infant mortality rate
The number of deaths of children (up to age one year) per 1000 births per year.

infection
Infection has occurred when viable organisms are present in the host's body.

innate behaviour
Behaviour that is inborn, controlled by genes and not influenced by learning.

innate releaser mechanism
A mechanism in an animal's brain that causes the animal to respond in an automatic way to a sign stimulus.

inter-sexual selection
Selection of a mate according to its qualities, such as bright colours or good health.

intra-sexual selection
Selection resulting from competition between members of the same sex for a mate or territory.

kinesis (plural, kineses)
Orienting behaviour in which an animal reduces its rate of movement or increases its rate of turning as the intensity of the stimulus increases.

latent period
The period of time between infection and the onset of antibody production.

learned behaviour
A long-lasting change in behaviour that is brought about by experience.

ligand
A molecule in a microbial cell wall or outer viral coat that binds with a receptor molecule on a cell membrane.

live vaccine (attenuated vaccine)
A vaccine containing weakened pathogens; the organisms cannot cause disease but do stimulate an immune response.

longitudinal study
A sample of people of the same age who are studied over a period of time.

luteinising hormone (LH)
A pituitary hormone causing ovulation by the mature Graafian follicle.

lysis
Rupturing of a cell and the loss of cellular contents.

male pill
A hormone-based pill, still under trial, taken by men as a contraceptive.

male pronucleus
The haploid set of chromosomes produced by a sperm after penetrating an oocyte, which fuses with the female pronucleus.

malignant tumour
A tumour of cancerous cells that continues to grow and may spread to other parts of the body.

memory B cell
B lymphocyte that continues to produce small amounts of antibody for years. B lymphocytes enable an individual to mount a rapid secondary response following subsequent exposure to a specific antigen.

menopause
The ending of the menstrual cycle, which occurs in women at about 50 years of age.

menstrual cycle
The periodic build up and breakdown of the endometrium coupled with the cyclic release of an oocyte by the ovary.

menstruation
The breakdown and loss of the endometrium at the start of each menstrual cycle.

metastasis
The spread of malignant tumour cells to other parts of the body.

microvilli
Folds of the cell membrane that greatly increase the area of exchange surfaces such as the placenta.

mineral deficiency disease
A disease that is caused by the dietary lack of a particular mineral.

monoclonal antibody
Pure antibody made from cloned lymphocyte cells fused to spleen cells.

morula
A stage of the embryo when it is a solid sphere of cells.

motivational state
The state of the nervous system or brain that makes an animal more or less likely to respond to a given stimulus.

myocardial infarction
The death of cardiac tissue, leading to a heart attack

myometrium
The outer layer of the uterus, made of muscle, and used during labour to give birth.

negative feedback
The method of regulation in which increasing the concentration of a substance in the body brings about changes that reduce its concentration.

negative phototaxis
Movement, using a taxis, away from the stimulus of light.

obese
Having excess body mass caused by a diet in which energy input exceeds the energy expenditure of the body.

oestrogen
A hormone made in the ovary and which helps to control the menstrual cycle.

operant conditioning
Conditioning in which a behaviour pattern is reinforced so that the behaviour becomes more frequent.

ovulation
The release of a secondary oocyte from a Graafian follicle of an ovary.

ovum (plural, ova)
An egg cell or oocyte.

pair bond
A behavioural attachment between male and female birds or mammals that is necessary for coordination of breeding.

pandemic
An epidemic that spreads across international boundaries.

parental investment
The time and energy put into producing and rearing offspring.

passive immunity
Natural: immunity acquired by receiving antibodies via the placenta or mother's milk.
Artificial: immunity resulting from the injection of ready-made antibodies in a vaccine.

pathogen
An organism that causes disease.

percentage growth rate (of a population)
The change in the number of people per unit time expressed as a percentage.

perivitelline space
The space between the oocyte membrane and the zona pellucida.

pheromone
A chemical substance that affects the behaviour of other members of the same animal species.

physiological age
The developmental age of an individual.

pituitary gland
The small outgrowth at the base of the brain that controls the endocrine system. The pituitary gland is under the influence of the hypothalamus.

pituitary growth hormone (PGH)
A hormone secreted by the pituitary gland in the brain. PGH stimulates the growth of tissues, by influencing the metabolism of proteins. (NB: also known as somatotrophic hormone or STH)

polar body
A tiny cell containing a haploid set of chromosomes but little cytoplasm, resulting from the meiotic division producing an oocyte.

polygamous
Having more than one mate.

polyspermy
More than one sperm penetrating an oocyte.

population growth
This is defined as (births plus immigration) minus (deaths plus emigration).

population pyramids
A visual representation of the age structure of a population.

population
The number of individuals of a species in a specified area.

positive feedback
The method of regulation in which increasing the concentration of a substance in the body brings about further increase its concentration.

positive phototaxis
Movement, using a taxis, towards the stimulus of light.

pre-pubertal period
The period of time between birth and puberty.

primary immune response
The response that occurs the first time an individual is exposed to a particular antigen.

primary oocyte
A developing egg cell part way through meiosis, temporarily stopped at prophase I.

primate
A member of the order of mammals that includes monkeys, apes and prosimians (i.e. primitive primates such as lemurs and tarsiers).

primordial follicle
An immature follicle, present in the ovary at birth, containing a primary oocyte.

progesterone
A hormone, made by the corpus luteum, that helps to control the menstrual cycle.

prostate gland
One of the accessory glands releasing fluids to form the semen.

puberty
The change from child to adult during which the child develops the secondary sexual characteristics found in adults.

radiographer
A person who carries out X-rays.

rate of natural increase
The change in the size of a population, as a percentage of the total population (usually a positive figure, but can be a negative value).

receptor protein
A protein on the surface or inside a target cell to which a hormone fits like a lock and key, initiating changes to the target cell.

reflex action
A rapid, automatic, innate response to a stimulus.

reflex escape response
A rapid, automatic, innate fleeing movement made by an animal in response to danger, often controlled by fast-conducting axons.

reinforcement
A reward or punishment that is used in operant conditioning.

reinforcer
A reward or punishment making a behaviour pattern more likely to be repeated.

screening
The process of looking for evidence of (potential) disease when the patient may not yet show any signs or symptoms.

secondary immune response
The response that occurs when an individual is exposed to a pathogen on a second or subsequent occasion. It is more rapid and produces greater levels of antibody than the primary immune response. The pathogen is usually destroyed before the development of signs and symptoms of the disease.

secondary oocyte
A developing egg cell part way through meiosis, temporarily stopped at metaphase II. The egg is at this stage when it is ovulated.

semen
The mixture of sperms and fluids that is ejaculated by a male.

seminal fluid
The fluid added to the sperms by the accessory glands to make the semen.

seminal vesicle
One of the accessory glands releasing fluids to form the semen.

senescence
The deterioration of body functions and the appearance of features associated with old age.

sensitive period
A time in the life cycle of an animal when it is especially likely to learn.

sensory nerve ending
The ends of sensory nerve cells in the skin. They are densest in areas such as the lips, mouth and finger-tips. They act as receptors, and send nerve impulses to the central nervous system when they receive a stimulus, such as touch.

sexual dimorphism
Differences in size or proportions in the male and female of a species.

sexual imprinting
Learning the characteristics of an animal's own species allowing it to select a suitable mate.

sign stimulus
A shape or colour that triggers species specific behaviour.

signs and symptoms
The signs that can be observed, and symptoms experienced by the sufferer, in association with a disease.

sinus (plural, sinuses)
A space (e.g. the space containing blood in the maternal part of the placenta).

somatic mutation
A mutation in body cells

spermatozoon (plural, spermatozoa)
Male gametes, abbreviated to sperms.

stem cell
A cell in an early embryo with the potential to develop into any type of adult tissue.

stimulus (plural, stimuli)
A change in an organism's environment; a stimulus might lead to a change in an animal's behaviour (a response).

survival curve
A graph that shows the number of survivors of a group of 10 000 people plotted against time.

syncytium
A group of cells with no cell membranes separating them.

taxis (plural, taxes)
Orienting behaviour in which an animal turns towards or away from a stimulus, such as light.

territory
Any defended area.

thrombosis
The condition in which there is a blood clot inside a blood vessel. Blood clots can block blood flow to an essential organ such as the heart or brain.

thrombus
A stationary blood clot in a blood vessel.

thyroid gland
An endocrine gland, located in the neck, that secretes the hormone thyroxine.

thyroid stimulating hormone (TSH)
A hormone secreted by the pituitary gland in the brain. TSH promotes the synthesis and release of thyroxine by the thyroid gland in the neck.

thyroxine
A hormone secreted by the thyroid gland. Thyroxine increases the metabolic rate of cells, and the mobilisation of glucose. Thyroxine affects differentiation and growth of tissues during childhood, influencing physical and mental development.

toxoid
A harmless form of a toxin.

trophoblast cell
A cell from the outer layer of the blastocyst. These cells, collectively known as the trophoblast, develop into the chorionic villi.

tubal ligation
A method of sterilising women by putting clips on the oviducts.

ultrasound
An imaging technique using high-frequency sound waves used to view internal organs or a developing fetus.

vaccine
A preparation that is derived from pathogens and that stimulates an immune response.

vasectomy
A method of sterilising men by removing a section of the vas deferens.

vitamin deficiency disease
A disease that is caused by a dietary lack of a particular vitamin.

warning coloration
The bright and conspicuous coloration found in many poisonous or distasteful animal species.

zona pellucida
The jelly coat around the oocyte.

Contents

INTRODUCTION

Choosing ten people who changed the world is a tricky task. Many more than just ten individuals have influenced the course of human history. However, only a few people are remembered today. In modern times, the most famous people in the world are often celebrities, sports stars or global leaders. But will they still be talked about in 100, or even 1000 years? To change the world, a person has to do something so extraordinary that it creates a legacy lasting far beyond their own death.

Astronomer Galileo Galilei is often called the 'Father of Modern Science.'

The ten people featured in this book came from different places and different times, and each one had a unique and enormous impact on the world that is still felt today. Some voyaged to undiscovered lands in the name of exploration or conquest. Others explored the world of science, to help humans on earth or explain the entire universe. Many tried to make people's lives better, while others sought to destroy whole cultures and civilisations. Some found that the smallest action could have a huge impact on society. Others showed that you do not need to be rich or powerful to make a difference – ordinary people can change the world, too.

Many people suffered while trying to change the world. They had to fight prejudice, discrimination and ignorance in their own lifetimes. Some world-changers lived their whole lives without understanding the significance of their actions. Others died wishing their impact had been greater in their lifetime.

While there is no standard model for someone who changes the world, there is one thing that unites the people in this book. Despite their differences, they all wanted to transform things and create something new. It is for this reason that all ten figures, for better or worse, changed the world we live in.

Florence Nightingale was a war heroine who helped found modern nursing.

Civil rights leader Nelson Mandela was South Africa's first black president.

ALEXANDER THE GREAT

STATS PANEL

Lived: 356 BC–323 BCE

World-changing moments: Founded over 70 cities and created an empire that stretched across three continents.

Titles: King of Macedonia, Persia, Asia and Pharaoh of Egypt.

Fact: Alexander named a city after his horse, Bucephalus.

In 334 BCE, Alexander the Great led 35,000 soldiers into Asia Minor (now part of Turkey) and began a ten-year campaign of conquest and control. With the 5-metre long 'sarissa' spear at the heart of his army, Alexander invaded a massive area of Persian territory stretching from modern-day Turkey to Iraq. It was the largest empire the ancient world had ever known. Alexander had vowed to crush the Greeks' oldest enemy, the Persians, since he was a boy. But Alexander himself was not Greek: he was Macedonian. During Alexander's childhood, Macedonia was a small kingdom that lay next to the great civilisation of Greece. The Greeks considered the Macedonians to be uncivilised barbarians. But Alexander's father, Philip II of Macedon, did something no Greek could achieve – he united all of Greece under his rule.

This ancient bust reveals Alexander's face to the modern world.

Alexander had been brought up in the Greek traditions and educated by the Greek philosopher Aristotle. He even modelled himself on the heroes of Greek myth and legend, Achilles and Hercules. So when his father was assassinated in 336 BCE, the newly crowned Alexander decided to conquer Persia. Alexander's army did not lose a single battle as it destroyed all Persian opposition. In 331 BCE, Alexander crushed the army of Persian King Darius III at the Battle of Gaugamela, and was proclaimed King of Asia. Alexander, however, yearned for new conquests. He marched his men through modern-day Afghanistan and Pakistan, but they would not follow him into India. Reluctantly, Alexander turned back. Alexander had conquered over two million square miles of land. Instead of dying a hero's death in battle, he died from a fever, possibly caused by a mosquito bite.

339 BCE: MACEDONIAN KING PHILIP UNITES GREECE UNDER HIS RULE .....

A painting of Alexander's victory over the Persians at the 333 BCE Battle of Issus.

Changing the world

Alexander's life was short, but he left a lasting legacy. In the former Persian Empire, he founded 70 new cities, created trade between the East and the West and spread the ideas of ancient Greece throughout his conqured lands. Today, this Greek influence can still be seen among the ruins of cities such as Alexandria in Egypt. Alexander allowed local customs to live alongside Greek culture,

helping his conquered subjects adapt to their new ruler. Yet, building a lasting civilisation proved beyond the King of Macedonia. After his death, Alexander's empire crumbled. Later, his accomplishments would serve as inspiration for many future warlords. Julius Caesar, Napoleon and Hitler all tried to copy Alexander's conquests, but none of them was able to match up to the man himself.

CHRISTOPHER COLUMBUS

Ten weeks after setting sail from Spain, Christopher Columbus reached the sandy shores of a new land. He assumed he was somewhere in Asia, known then as the 'Indies', and called the native inhabitants 'Indians'. But Columbus was nowhere near Asia. Instead, he had accidently reached the islands today called the Bahamas, in the Caribbean. Over the next 12 years, Columbus made three return voyages to various islands of the Caribbean, as well as central America and Venezuela, on behalf of the Spanish crown. Columbus was one of the first Europeans to visit these places and he later became known as 'the man who discovered America'.

Columbus was born in the Italian city of Genoa and brought up during the Age of Discovery when Europeans were searching for new sea routes to Asia in their hunt for trade, wealth and land. Wishing to join these explorers, Columbus persuaded Queen Isabella and King Ferdinand of Spain to finance a voyage to Asia. Most European captains sailed around Africa to Asia, but Columbus attempted to reach it by crossing the Atlantic Ocean.

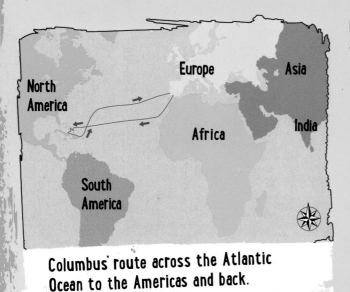

Columbus' route across the Atlantic Ocean to the Americas and back.

In 1492, Columbus set out with the fleet of ships that would carry him to the Americas: the Santa Maria, the Pinta and the Niña. Columbus claimed every territory he visited for Spain and was made 'Admiral of the Seven Seas' and 'Viceroy of the Indies'. However, reports reached the Spanish court of Columbus' brutal treatment of the native inhabitants and he was removed from his position as viceroy. Columbus and his heirs tried to sue the Spanish court for money made from his discovered territories, but were unsuccessful. He died a disappointed man in 1506.

1492: CHRISTOPHER COLUMBUS DISCOVERS THE AMERICAS FOR EUROPE .....

Lived: 1451-1506

World-changing moments: Discovered the Americas.

Titles: Admiral of the Seven Seas and Viceroy of the Indies.

Fact: Columbus convinced the Spanish court of the riches to be found in the new world by bringing back gifts of gold, spices, parrots and human captives.

Christopher Columbus is shown landing in the Bahamas for the first time.

Changing the world

Although the Vikings had visited the Americas several centuries earlier. Columbus was the first explorer to open up the continents to Europe. This meant an exchange between the two worlds of cultures. people. plants. animals and diseases. This world–changing moment had some devastating consequences for the native peoples of the Americas. The Taino Indians of Hispaniola suffered enslavement and brutality at the hands of Columbus. while others were wiped out by European smallpox. These devastating developments continued in the centuries to come as European nations conqured more of the Americas. with the Aztecs. Incas and Native Americans all suffering similar fates. Columbus himself would never have dreamed of this legacy. He went to the grave still believing the lands he had discovered were in Asia.

GALILEO GALILEI

Galileo Galilei was an Italian astronomer who was imprisoned for his revolutionary ideas about the solar system. Educated as a mathematician, Galileo's life took a dramatic turn in 1609 when he heard about the invention of the telescope. He set about building a superior model so he could observe the stars above. Soon afterwards, Galileo discovered four moons travelling around Jupiter. This lead him to believe that the planets were also travelling around a large body – the sun. The theory that the sun, and not the earth, was at the centre of the universe had already been put forward by Polish astronomer Nicholas Copernicus. But this view, now also supported by Galileo, contradicted the teachings of the Catholic Church. The Church accused Galileo of heresy – a serious crime that could result in a death sentence.

STATS PANEL

Lived: 1564-1642

World-changing moments: Spoke out against the beliefs of the Chruch – separating science and religion.

Scientific fields: Astronomy, physics and mathematics.

Fact: Galileo was going to become a priest before his father convinced him to study medicine. He then swapped to mathematics.

Galileo Galilei explains his theory about the Earth's place in the universe.

Galileo was a gentleman of the court in Florence, Italy, and managed to escape punishment for supporting Copernicus' theory. However, in 1616, the Church forbade Galileo from speaking out about it. For years he remained silent, but the theory emerged once more in Galileo's book *Dialogue on the Two Chief World Systems* in 1632. Galileo was put on trial by the Roman Catholic Church and found guilty of heresy. He was sentenced to life imprisonment, which he spent under house arrest in a villa in Arcetri, near Florence. From here, Galileo continued his investigations into physics and astronomy, and made new discoveries in both areas. In his later life, Galileo suffered from various illnesses and he was completely blind at the time of his death.

1609: GALILEO OBSERVES THE SKY THROUGH HIS CUSTOM-MADE TELESCOPE

Galileo observes the sky through his custom-made telescope.

Changing the world

Galileo's contributions revolutionised the world of science and religion. As an astronomer, Galileo made a series of important discoveries with his telescope: he found that the moon is rough not smooth, and that there were many more stars in the Milky Way than previously believed. Galileo was interested in other scientific fields, too, and his work later contributed to Isaac Newton's Laws of Motion.

He also helped develop the 'scientific method', the process used in scientific investigation. However, it is for his conflict with the Church that Galileo is best remembered. He refused to accept the Church's ideas unless they could be scientifically proven. He began the separation of religious belief and scientific thought. It is also one of the reasons Galileo is known today as the 'Father of Modern Science'.

MARIE CURIE

STATS PANEL

Lived: 1867-1934

World-changing moments: Discovered the chemical elements radium and polonium.

Awards: The Nobel Prize for Physics and the Nobel Prize for Chemistry.

Fact: Curie was the first women to win a Nobel Prize, the only woman to win the Prize in two fields, and the only person to win in multiple sciences.

Marie Curie was a world-changing scientist who had to fight to be educated and work in a field dominated by men. Curie was born Marie Sklodowska in Poland at a time when women were banned from higher education. To escape this policy, Curie attended a secret University in Warsaw before moving to the Sorbonne University in Paris, France. Curie had little knowledge of French and barely any money. She lived in a small attic room where she had to wear several layers of clothing to keep warm. Curie often could not afford to buy food and sometimes fainted from hunger. She also faced discrimination at the university for being a woman. Curie did not give up despite these obstacles and obtained degrees in both maths and physics.

> 66 Nothing in life is to be feared. it is only to be understood. Now is the time to understand more, so that we may fear less. 99
>
> Marie Curie

Marie helped change attitudes towards female scientists.

It was at university that Marie met and married physicist Pierre Curie. Together, the pair discovered the chemical elements radium and polonium and coined the term 'radioactivity'. This won the Curies the Nobel Prize for Physics in 1903. It was the first time a woman had been awarded the prize. But tragedy struck when Pierre was knocked down and killed by a horse-drawn carriage. Left to bring up their two daughters alone, Curie continued her work and was made professor at the University of Paris in her husband's place. Curie went on to isolate the element pure radium, a groundbreaking achievement which won her the Nobel Prize for Chemistry in 1911. Her research would lead to the use of radiotherapy in medicine, but it was also took its toll on Curie's health. In 1934 she died from leukaemia, caused by her long-term exposure to radiation.

1903: CURIE WINS THE NOBEL PRIZE AND COINS THE PHRASE 'RADIOACTIVITY' .....

Changing the world

Marie Curie changed the world through her scientific discoveries and by changing attitudes towards female scientists. Her research into radioactivity provided a major scientific breakthrough that led to the use of radiation therapy to fight cancer and helped later scientists better understand the atom. It also enabled Curie to develop an X-ray service for war-time doctors who used it to locate shrapnel and broken bones in patients before surgery.

Curie even drove the vans containing portable X-ray units to hospitals on the frontline during the First World War (1914–1918). Despite her accomplishments, Curie had to fight for her career – first to be educated and then to prove her worth as a scientist. It was through her extraordinary academic ability that Curie was able to transform attitudes towards women in the scientific community, and help secure their place within it as equal partners.

FLORENCE NIGHTINGALE

Florence Nightingale was born into a wealthy family during the Industrial Revolution and lived a privileged life. While many English people struggled to survive working in factories and coal mines, Nightingale's time was spent socialising, entertaining and taking holidays. However, she became deeply unsatisfied. Her ambition was to become a nurse, but her parents were against the idea. Nurses at that time were untrained and were considered little more than servants by the upper classes. But Nightingale went against her parents' wishes and became a nurse despite their objections. At 33 years old, her first job was running a private hospital in London. She would go on to found the profession of modern nursing and become a British war heroine.

STATS PANEL

Lived: 1820-1910

World-changing moments: Changed people's attitdues towards nursing and nurses.

Title: Nightingale became famous in Britain as 'The Lady with the Lamp'.

Fact: Nightingal's title 'The Lady with the Lamp' was coined after a *Times* newspaper illustration showed her tending patients at night by lamplight.

A bird's-eye view of the British cavalry camp during the Crimean War.

Nightingale became famous after she led a team of nurses to tend to British soldiers during the Crimean War. She was horrified at the conditions in the army hospital in Scutari in modern-day Turkey. Wounded men lay unattended on the floor of filthy, overcrowded wards, infested with rats, fleas and lice. Nightingale transformed the hospital by introducing order, sanitation and patient care. She worked for twenty hours a day and tended to the soldiers late into the night, which earned her the title 'Lady with the Lamp'. Newspaper reports about Nightingale made her famous in Britain before she even returned home. Yet, Nightingale's own health suffered back in England and she spent the last 50 years of her life confined to her house. Despite this, she opened the Nightingale School for Nurses in London in 1860, the first of its kind in the world.

1860: NIGHTINGALE OPENS THE FIRST PROFESSIONAL SCHOOL FOR NURSES ..... PRESENT DAY:

Florence Nightingale was an important figure in the struggle for equal rights for women.

Changing the world

Florence Nightingale changed the world of nursing and subsequently changed commonly held views about women in the workplace. From the moment Nightingale entered the nursing profession, it was her dream to train nurses. Nightingale's school for nurses and her textbook *Notes on Nursing* set a benchmark for modern nursing. The 'Nightingale nurses' quickly gained a reputation for the excellence that Nightingale's name is still associated with today.

It is for this reason that Nightingale is considered to have laid the foundations for modern nursing. But Nightingale's work also marked an important moment for women's rights. Although she was expected to stay at home and raise a family, Nightingale showed that women were capable of taking up and succeeding in professional occupations that had previously been closed to them.

ADOLF HITLER

"The personification of the devil as the symbol of all evil assumes the living shape of the Jew."

These are Adolf Hitler's words from *Mein Kampf* (My Struggle), a memoir which outlined his hatred of Jewish people and his plans for a new world order. It was with these plans in mind that Hitler caused the outbreak of the Second World War in 1939 and the deaths of millions, including the extermination of six million Jews. When Hitler wrote *Mein Kampf* in 1925, he was an almost unknown figure. He was also in jail, having been imprisoned in 1923 for trying to lead an armed uprising against the German government. At that time, Hitler was the leader of the fascist National Socialist German Workers' Party, also known as the Nazi party. He had joined the Munich-based party after fighting as a soldier in the First World War.

One of Hitler's speeches rallying the German people to his cause.

The Auschwitz–Birkenau concentration camp where over 1 million people were executed.

After being released from prison, Hitler set about putting his plans into action. By 1933, the Nazi party was in government and Hitler was the country's leader. He immediately began strengthening Germany's military and introduced a series of anti-Jewish laws, which were enforced with violence. Hitler's invasion of Poland in 1939 started the Second World War. As his army conquered more countries across Europe, Jews and other people considered racially inferior were rounded up and killed. After two successful years of war, the tide turned agains Hitler. By April 1945, Allied troops had won back all of Europe's conquered territories and were attacking Berlin. On 30 April, Hitler committed suicide in his bunker.

Changing the world

Hitler showed the power of one man in leading a nation to commit war and mass-murder on a world-changing scale. It is widely accepted that Hitler's beliefs led to the deaths of 29 million soldiers and civilians during the Second World War. However, the true extent of the war was not known until after the fighting had finished. Across German occupied territories, over 300 concentration camps had been created to execute people or work them to death. The mass extermination of Jews, Roma people, homosexuals, communists and anyone else Hitler's government considered its enemy, became known as the Holocaust. The total number of people executed by the Nazis is thought to be over 19 million. It is this unimaginable horror, alongside the destruction of much of Europe, that makes up Hitler's terrible legacy.

ALAN TURING

Alan Turing was often described as an eccentric: he wore a gas mask in summer to prevent hay fever and chained his mug to a radiator at work so it would not be stolen. He was also a genius who helped the Allies win the Second World War by cracking the Nazi code-making machine, Enigma. Turing was hired to work at Bletchley Park, the British government's code-breaking headquarters, after studying mathematics at Cambridge University. At Bletchley Park, Turing developed the 'bombe', a device that could decipher Enigma's coded messages. The Nazis considered their Enigma code-making machine impossible to crack and it was used for all high-level operations by the German army, air force and navy. Cracking Enigma created a breakthrough moment in the war that enabled the Allies to unscramble German messages and ultimately defeat them in battle.

STATS PANEL

Lived: 1912-1954

World-changing moments: Cracked the Enigma code.

Titles: Called the Father of Computer Science.

Fact: Turing was a talented long-distance runner who would often run the 40 miles (64 kilometres) from Bletchley Park to London for top-secret wartime meetings.

Alan Turing not only cracked the Nazi Enigma machine but also helped develop modern computing.

The Enigma code-making machine used by the Nazis.

After the war, Turing helped found modern computer science through his work at the National Physical Laboratory and Manchester University. Turing's ACE (Automatic Computing Engine) and the Manchester Mark 1 were among the first programmable computers ever invented. Turing also laid the groundwork for artificial intelligence and his 'Turing test' is still used today to tell whether an internet user is a computer or a real person. Despite his valuable work, Turing's world was turned upside down in 1952. This was the year he was arrested and found guilty of homosexuality, a punishable crime in Britain until 1967. Although he escaped going to jail, Turing's security clearance was revoked, meaning he could no longer work for the government. Driven to despair, Turing committed suicide in 1954.

1945: TURING AWARDED AN OBE FOR CRACKING ENIGMA ..... 1952: TURING PROSECUTED

> 66 Turing deserves to be remembered and recognised for his fantastic contribution to the war effort and his legacy to science. A pardon from the Queen is a fitting tribute to an exceptional man. 99
>
> **Alan Grayling, Secretary of State for Justice, 2013**

Changing the world

Turing's work cracking Enigma saved countless lives during the Second World War and was described by Prime Minister Winston Churchill as "The single biggest contribution to Allied victory in the war against Nazi Germany." It was for his world-changing wartime work that Turing was awarded an OBE (Order of the British Empire). It is a mark of his genius that Turing is also remembered for his contributions to modern computing. Turing has often been called "The Father of Computer Science" and his name is still celebrated through awards, computer programmes, stamps and plaques. Despite his world-changing contributions, Turing's life was cut short in the most tragic of circumstances. In 2009, the British Government apologised for the "appalling" way Turing had been treated for being homosexual and in 2013 the Queen granted him a posthumous pardon.

ROSA PARKS

On an ordinary December day in 1955, 42-year-old seamstress Rosa Parks boarded the bus to get home. Parks was an ordinary working woman who was unknown to the world – but she was about to change history. Parks sat down as usual on the bus in the seats reserved for black people. In Montgomery, Alabama, USA at that time, buses were segregated into a white section at the front and a black section at the back. But as the bus filled up, the driver asked Parks to stand so a white man could take her seat. She replied that she would not. The driver called the police and Parks was arrested and later fined by the courts. In the 1950s, Alabama was one of several southern American states that had segregated public spaces. This meant black people were banned from using the same facilities as white people, including park benches, cafés, toilets and libraries.

Rosa Parks' small protest helped ignite the American civil rights movement.

Rosa Parks sits by a white man after the government rules segregation on buses is illegal.

News of Parks' protest on the bus sparked a 381-day boycott of Montgomery buses by black people. The boycott was organised by a young minister called Martin Luther King Jr., who went on to become a famous leader of the American civil rights movement. The boycott worked. In 1956, the U.S. Supreme Court ruled that the segregation law was unconstitutional and the black section on Montgomery buses was abolished. The ruling was a great victory in the fight for equal rights for black people that sparked non-violent protests across America. Parks, however, was fired from her job as a result of her protest. She later worked as a secretary for a congressman and co-founded the Rosa and Raymond Parks Institute for Self Development, which provided career training for young people.

1955: ROSA PARKS REFUSES TO STAND FOR A WHITE MAN ON A SEGREGATED BUS .....

> 66 I don't know why I wasn't, but I didn't feel afraid. I had decided that I would have to know once and for all what rights I had as a human being and a citizen, even in Montgomery, Alabama. 99
>
> **Rosa Parks (1956)**

Changing the world

To the eyes of the world, Rosa Parks was an ordinary woman living an ordinary life before that fateful day in 1955. But through her quiet, courageous action she changed the course of history. In the 1950s, many black people in the American South began to fight the policy of segregation. When Parks was a girl, she watched the schoolbus for white children only drive past as she walked the several miles to school. It was because of

Parks' protest on the bus years later and the civil rights protests which followed, that black people achieved legal equality in the 1960s. This gave them the same rights to vote, live, be educated and sit in the same places as white people. It is for this reason, Rosa Parks, who seemed like an unremarkable seamstress from Montgomery, is today remembered as the 'Mother of the Civil Rights Movement'.

OSAMA BIN LADEN

Osama bin Laden was the leader of Islamist terror group al-Qaeda and the mastermind behind the terror attacks on 11 September 2001 in New York, USA. Bin Laden was one of 50 children born to self-made billionaire Muhammad bin Laden and was brought up in Saudi Arabia. At university, bin Laden developed radical Islamist ideas and a strong hatred of Western governments. In 1979, he travelled to Afghanistan to gain military training and help the local Taliban fight against the Soviet Union's occupation of the country. The Taliban is a group of Islamic fundamentalists made up of Afghanistani tribal warriors. Bin Laden raised funds for the Taliban and also created a military network of Arab fighters that would later become al-Qaeda. Al-Qaeda became known for its tactics of suicide attacks and the bombing of buildings.

STATS PANEL

Lived: 1957-2011

World-changing moments: Masterminded the 9/11 terror attacks on the USA.

Fact: From 2001-2011 bin Laden was at the top of the USA's most wanted list, with a US$25 million bounty on his head.

Osama Bin Laden, the mastermind behind the 9/11 attacks on the USA.

One of the World Trade Center towers after a plane has been flown into it.

In the 1990s, al-Qaeda began a series of terrorist attacks against the USA that included the bombing of its embassies in Kenya and Nairobi. In 1998, bin Laden called for all Muslims to kill Americans and their allies wherever they could. This led to al-Qaeda hijacking four planes for its 9/11 attacks. The attacks resulted in the destruction of New York's two World Trade Center skyscrapers and the deaths of nearly 3,000 people. In response, the United States ordered the military overthrow of the Taliban in Afghanistan. In 2011, bin Laden was found hiding in a walled compound in the city of Abbottabad, Pakistan. He was killed during a US raid on the compound and his body buried at sea.

2001: OSAMA BIN LADEN LAUNCHES HIS 9/11 ATTACKS ON AMERICA .....

Changing the world

Osama bin Laden's terrorist actions have had a massive impact on modern civilisation. The main objective of bin Laden's al-Qaeda was to end the presence of the USA and other Western countries in Muslim countries. But instead, the 9/11 attacks led the USA and other Western countries to begin to fight a global war on terror that continues today. Western attacks within Muslim territories has led to the deaths of thousands of military personnel and civilians alike. Everyday lives have also been affected, as fears of further terrorist attacks and heightened security in public places have become regular features of modern life in Western cities. The war continues against al-Qaeda and other terror groups that include the Islamic State of Iraq and Syria (ISIS). As the world still grapples with the issues caused by bin Laden and al-Qaeda, it is difficult to predict how severe and long-lasting his legacy will be.

NELSON MANDELA

In 1964, Nelson Mandela was found guilty of treason and narrowly escaped the death penalty. Instead, the South African civil rights leader was sentenced to 27 years in prison. The government thought Mandela's imprisonment would signal the end of his fight against its policy of apartheid. Apartheid was South Africa's segregation of races, which banned black people from voting, forced them to live in special 'black' areas with poor conditions. However, the government was wrong to think it could silence Mandela. Instead, his imprisonment brought Mandela attention and support from around the world and put pressure on South Africa's government to abandon apartheid.

STATS PANEL

Lived: 1918-2013

Awards: Mandela won more than 250 awards during his lifetime, including the 1993 Nobel Peace Prize.

Titles: President to South Africa 1994-99 and called the 'Founding Father of Democracy.'

Fact: Mandela was a member of the tribal Xhosa-speaking Tembu people and was often called by his clan name, Madiba.

Nelson Mandela leaves prison for the first time after 27 years.

Rolihlahla 'Nelson' Mandela first began his struggle against oppression as a law student in Johannesburg. Here, he joined the African National Congress (ANC), a party that fought for the rights of black people. Apartheid was introduced in 1948 and the ANC battled against the racist policy through non-violent protests. Following the shootings of 69 anti-apartheid protestors during a demonstration in the black township of Sharpeville in 1960, the government declared a state of emergency and banned the ANC. In response, the ANC abandoned its policy of non-violence and Mandela sought out military support for the party in other countries. When he returned, Mandela was arrested and imprisoned. Twenty-seven years later, the South African government bowed to international pressure by releasing Mandela. Mandela then worked with the government to abolish apartheid in South Africa. He became the country's first black president in 1993 and died a free man twenty years later.

1964: NELSON MANDELA IS BRANDED A TERRORIST AND IMPRISONED FOR 27 YEARS

Nelson Mandela used his fame to promote important causes until his death.

Changing the world

At the time of his death, Nelson Mandela was one of the most famous figures in the world. In South Africa, he was known simply as the 'Founding Father of Democracy'. For those living under apartheid, Mandela was their liberator and champion. Mandela's imprisonment did little to stop his influence and instead brought widespread international condemnation. After his release in 1993, Mandela inspired the world by working with the government that had imprisoned him to create a new South Africa with equal rights for all. His work with then president F. W. de Klerk to achieve this aim made both men co-winners of the 1993 Nobel Peace Prize. After standing down as president, Mandela continued to use his fame to promote good causes, including raising awareness of the deadly AIDS/HIV virus. Today, Mandela's image is instantly recognisable as a symbol for the fight against oppression and a reminder that people have the ability to change the course of human history for the better.

10 OTHER PEOPLE THAT CHANGED THE WORLD

1. Leonardo da Vinci (1452–1519)

An artist, engineer, scientist and inventor who was considered one of the most creative minds of the Italian Renaissance. Painter of the famous *Mona Lisa*, da Vinci also drew up plans for a bicycle and helicopter 500 years before they were invented.

2. William Shakespeare (1564–1616)

An English writer, poet and actor considered by many to be the greatest playwright of all time. His work may be over 400 years old, but it is still taught in schools today and many of his phrases are used in everyday language.

3. Abraham Pineo Gesner (1797–1864)

A Canadian geologist and doctor who invented kerosene, a liquid fuel that was originally used in lamps. Today, over 1 billion barrels of kerosene are used worldwide every day for cooking, heating and in jet fuel for aeroplanes.

4. Guglielmo Marconi (1874–1937)

An Italian inventor, engineer, physicist and winner of the 1909 Nobel Prize for Physics. Marconi is best known for inventing the wireless telegraph, which paved the way for all modern radio

5. Mahatma Gandhi (1869–1948)

The leader of the nationalism movement in India, who used non-violent protest to bring about independence from British rule from 1858 to 1947. Gandhi inspired civil rights movements around the world and is often called the 'Father of India'.

6. Joseph Stalin (1879-1953)

A Russian dictator who ruled over the Soviet Union for quarter of a century using a regime of terror. Stalin helped defeat the Nazis and turned the Soviet Union into an industrial country. However, he also caused the deaths of tens of millions of his own people.

7. Edmund Hillary (1919-2008)

A New Zealand explorer who became the first man to climb Mount Everest, the highest mountain in the world. He later created the Himalayan Trust to build schools, hospitals and airfields for the Nepalese people.

9. Tim Berners-Lee (1955-)

A British computer scientist who invented the World Wide Web. He also developed the world's first web browser and website, which explains the World Wide Web: http://info.cern.ch

8. Rosalind Franklin (1920-1958)

A British chemist who helped discover the structure of deoxyribonucleic acid (DNA), the building block of life in all living things. Franklin's contribution to the discovery was not recognised until after her death at 37 years old.

10. Mark Zuckerberg (1984-)

An American internet entrepreneur who co-founded the social networking website Facebook. The website helped usher in a new era of online social media. Zuckerberg was named one of the wealthiest and most influential people in the world in 2010 by *Time* magazine.

TIMELINE

334 BCE
Alexander the Great lands in Asia Minor and begins his overthrow of the Persian Empire.

476 CE
Roman Emperor Romulus Augustus is deposed by Germanic chieftain Odoacer, marking the beginning of the end of the Western Roman Empire.

1066
Anglo-Saxon King Harold II is defeated by Duke William II of Normandy at the Battle of Hastings, which results in the Norman conquest of England.

1914
Austrian archduke Franz Ferdinand is assassinated in Sarajevo, leading to the outbreak of the First World War in Europe.

1903
Marie Curie is awarded the Nobel prize for Physics, making her the first woman to win the prize.

1865
The American Civil War between the northern and southern states ends. resulting in the abolition of slavery in the United States.

1917
The Bolsheviks lead an armed uprising in Petrograd as part of the Russian Revolution that deposes the Tsar and establishes communist rule.

1939
German Chancellor Adolf Hitler orders the invasion of Poland, marking the start of the Second World War.

2001
Islamic terrorist group al-Qaeda, led by Osama bin Laden, hijack four planes in an attack on the Unites States of America that includes the destruction of the New York World Trade Center.

1990
Civil rights leader Nelson Mandela is released from prison, resulting in the end of South Africa's apartheid system.

1969
American astronaut Neil Armstrong becomes the first human to set foot on the moon, signalling the end of the Space Race between the Soviet Union and the United States.

1215
The Feudal Barons of England sign the Magna Carter with King John, limiting the king's powers and setting out a new law of the land.

1492
Christopher Columbus lands in the Bahamas and begins his discovery of the Americas for Spain.

1633
Galileo Galilei is sentenced to life imprisonment by the Catholic Church for supporting Copernicus' theory that the earth is not at the centre of the universe.

1854
Florence Nightingale leads a team of nurses to aid British soldiers wounded in the Crimean War, where she becomes known as 'The Lady with the Lamp.'

1766
Thirteen American colonies sign the Declaration of Independence which proclaims they are no longer part of the British Empire, but instead a new 'United States of America'.

1940
Alan Turing's 'bombe' device cracks the Nazi code-making Enigma machine, enabling the Allies to understand secret German war messages.

1945
The United States drops an atomic bomb on the Japanese city of Hiroshima, leading to the country's surrender and the final end of the Second World War.

1989
The Berlin Wall separating western Germany from the socialist German Democratic Republic in the east is torn down, resulting in German reunification.

1955
Rosa Parks becomes a famous civil rights activist when she refuses to give up her seat in the 'black people's' section of a public bus to a white man.

GLOSSARY

abolish To formally put an end to something.

Age of Discovery A period between the fifteenth and seventeenth centuries when Europeans began exploring the world for goods, trade routes and land.

AIDS A disease caused by a virus.

Allies The countries, including France and Britain, that were allies during the first and second world wars.

Apartheid The former political system in South Africa in which only white people had full political rights and other people, especially black people, were forced to live in separate areas and use their own schools, hospitals etc.

artificial intelligence Computers and machines behaving with human-like intelligence.

assassinate To murder an important person, especially for political reasons.

atom The smallest unit of matter.

barbarian A member of a people not belonging to one of the great civilisations.

boycott The refusal to use or buy goods from someone as a form of protest.

civil rights movement The national effort made by black people and their supporters in the 1950s and 1960s to eliminate segregation and gain equal rights.

colonise Sending settlers to a new place and taking control of it.

conquest To take control of a place or people.

contradict To say the opposite of a statement made by someone else.

Crimean War A 1853–1856 conflict fought between Russia and an alliance of Britain, France, the Ottoman Empire and Sardinia.

decipher To understand and convert a coded message into normal language.

democracy A system of government by which members elect representatives.

depose To remove forcibly.

discriminate To treat a person, or group of people, differently from the rest.

enslave To make someone into a slave, who works for their master without being paid.

execute To kill somebody, especially as a punishment for breaking the law.

exterminate To destroy something completely.

fascism An extreme system of government led by a dictator.

heresy A belief or idea that contradicts the teaching of the Church.

hijack To illegally seize control of a vehicle.

HIV A virus that damages the immune system (the body's defences against disease) so that the sufferer catches diseases easily. If no treatment is given, the HIV infection causes AIDS.

Holocaust The mass murder of Jews and other people by Nazis during the Second World War.

Industrial Revolution The fast development

of Industry that began in Britain in the late eighteenth century.

invade When an army from one country enters another country by force to take control of it.

Islam A world religion. Followers of Islam are called Muslims.

leukaemia A life-threatening disease caused by an increase in human white blood cells.

liberate To provide freedom.

Milky Way The galaxy that contains our Solar System.

Nobel Prize An annual award given for scientific or cultural advances.

posthumous Given after a person's death.

prejudice A preconceived opinion that is not based on reason or actual experience.

regime A system of government, often one that is extreme.

Renaissance The revival of classic art and literature in the fourteenth to sixteenth centuries.

reunification When something is put it back together.

revolutionary To cause a dramatic change in something, such as a long-held belief.

Russian Revolution A series of revolutions in the Russian Empire during 1917.

sanitation Keeping a place clean from dirt, disease and infection by removing waste and providing clean water.

segregation The enforced separation of a race, class, or ethnic group.

smallpox A disease caused by a virus that causes fever and postules that eventually scar.

territory An area of land under a ruler or state.

treason The act of betraying your country.

unconstitutional Something that is not in agreement with political rules of the time.

uprising An act of rebellion.

viceroy A ruler exercising authority in a colony on behalf of a king or queen.

FURTHER INFORMATION

Books

Inspirational Lives: Nelson Mandela,
Kay Barnham, Wayland (2014)

Black History Makers: Campaigners,
Debbie Foy, Wayland (2014)

Who's Who in Science and Technology,
Bob Fawke, Wayland (2014)

Websites

100 historical people who changed the world:
www.biographyonline.net/people/people-who-changed-world.html

Time *magazine's most 100 influential people:*
http://content.time.com/time/specials/
packages/0,28757,2020772,00.html

Famous historical people from around the world:
www.kidinfo.com/american_history/famous_
historical_people.htm

Places to visit

The Florence Nightingale Museum: 2 Lambeth
Palace Road, London SE1 7EW

Bletchley Park: Sherwood Drive, Bletchley,
Milton Keynes, MK3 6EB

Imperial War Museum North: the Quays,
Trafford Wharf Road, Manchester, M17 1TZ

INDEX